AF332944

T. R. Tölle   J. Schadrack
W. Zieglgänsberger (Eds.)

Immediate-Early Genes in the Central Nervous System

With 42 Figures

Springer

Dr. Dr. T. R. Tölle
Dr. J. Schadrack
Prof. Dr. W. Zieglgänsberger

Max-Planck-Institut für Psychiatrie
Klinisches Institut
Klinische Neuropharmakologie
Kraepelinstr. 2–10
D-80804 München, Germany

ISBN 3-540-58962-7 Springer-Verlag Berlin Heidelberg New York

CIP data applied for

Typesetting: Camera ready copy from the editors
SPIN 10473645   25/3136 - 5 4 3 2 1 0 - Printed on acid-free paper

Preface

Immediate-early genes are believed to be involved in the neuron's ability to convert short-term synaptic stimulation into long-lasting responses and thus contribute to the adaptive alterations involved in neuronal plasticity. Cellular immediate-early genes share a close structural homology with some viral oncogenes. Recent advances in cellular biology have identified the activation and deactivation of immediate-early genes as molecular mechanisms to control regulated and deregulated growth, cellular differentiation and development. In this view immediate-early genes may function as third messengers in a stimulus-transcription cascade transferring extracellular information into changes in target-gene transcription, thereby changing the phenotype of neurons.

Immediate-Early Genes in the Central Nervous System provides a comprehensive up-to-date overview of current methodology in the research of immediate-early genes and includes a wide range of neurobiological topics, such as regeneration, memory formation, epilepsia and nociception. The contributors to this book have been selected from among the leading experts in their field of research.

T.R. TÖLLE
J. SCHADRACK
W. ZIEGLGÄNSBERGER

Immediate-early genes - how immediate and why early?

G.I. Evan
Imperial Cancer Research Fund Laboratories, 44, Lincolns Inn Fields, London
WC2A 3PX, UK

Introduction

The prerequisite for multicellularity is the effective integration of component cells in the formation and maintenance of complex tissue architectures. As a consequence, metazoan cells require sophisticated molecular machinery with which to communicate with their environment. Over the past decade there have been substantial advances in our understanding of the molecular processes by which cells receive, transduce and generate signals. The molecules that comprise such signal transduction pathways include a bewilderingly diverse set of functions, reflecting the diverse consequences that signals generate in differing cell types. However, one of the most intriguing general findings is that many components of signal transduction pathways can function as oncogenes if their activity becomes deregulated or inappropriately activated: that is, they trigger inappropriate and unrestrained cell proliferation leading to neoplastic transformation. This in turn indicates that cell proliferation is under the control of signalling pathways and, therefore, regulated by the availability of external mitogenic signals. The importance of signal transduction pathways in carcinogenesis has largely driven the recent enormous advances in our understanding of the molecular basis of these processes.

Cell proliferation is, however, only one of the possible cellular fates that can be triggered by extracellular signals. Signalling pathways regulate growth arrest, differentiation, programmed cell death (apoptosis) as well as a variety of other long term adaptive responses, such as the establishment of cell-cell communication and contact and more arcane phenomena such as memory. It is, perhaps, especially intriguing that many of the same signalling pathways are used to regulate each of these very different cellular responses, often within a single cell type. Thus, there do not seem to be specific signalling systems for specific cellular functions: rather, a generic signalling machinery is used to transduce diverse signals leading to a wide range of cellular responses (Fig. 1). Exactly where within the signalling machinery the required specificity resides remains largely a mystery. It seems likely, however, that signalling pathways in biology are not linear and unitary but highly iterative and parallel - more akin to networks than wires.

In this introduction, I will concentrate primarily on those immediate-early re-

Cell Fates

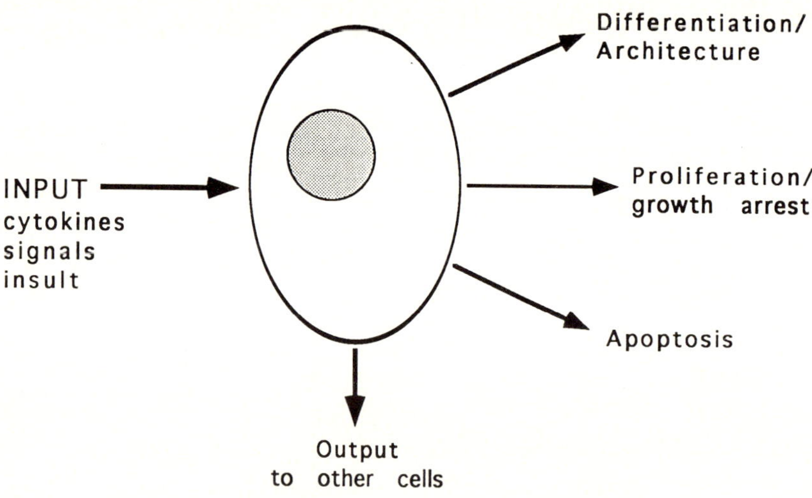

Fig. 1. Diagrammatic representation of the cell as a processor of information.

sponses whose products regulate gene expression and use them to illustrate some essential points. More detailed accounts of individual immediate-early genes, their products, their functions, and their biologies will be provided in subsequent chapters.

Immediate-early genes in mitogenesis

Immediate-early (IE) genetic responses were first described during the mitogenic induction of quiescent mammalian lymphocytes and fibroblasts. In the absence of mitogenic growth factors, most mammalian cells enter a quiescent state that is usually typified by a G1-like content of DNA and is often termed G0. Quiescence appears to be a fundamentally different state from any encountered during the cell cycle of exponentially proliferating cells. The classic G1/S cell cycle checkpoint, regulated by various cell cycle factors such as $p105^{RB}$ phosphorylation, cyclin E-Cdk4 activity and p53, and thought to be equivalent to START in yeast, is not the same as G0. Start maps temporally to about 2 hours prior to onset of S phase. In contrast, G0 cells take considerably longer to enter S phase following mitogenic stimulation and the transition from G0 into G1 is accompanied by a diverse pattern of genetic changes involving expression of immediate-early (IE) and delayed-early genes. Mitogenic induction of IE genes does not require *de novo* protein synthesis implying that all machinery required for their expression pre-exists within quiescent cells.

In the early 1980s, several studies demonstrated that certain immediate-early growth response genes were known proto-oncogenes - that is, genes whose deregulated activity triggered inappropriate cell proliferation. Both c-*fos* and c-*myc* were identified in this context (Cochran et al. 1984; Greenberg and Ziff 1984; Hogquist et al. 1991; Kelly et al. 1983; Muller et al. 1984). Subsequent to this, differential screening and subtractive cDNA cloning approaches have revealed there to be a large repertoire of immediate-early mitogen response genes (see below) (Almendral et al. 1988; Lau and Nathans. 1985). Comparisons of the repertoire of IE genes in different cell types such as fibroblasts, lymphocytes and liver cells (Mohn et al. 1991) shows that many IE genes are common to mitogenesis in many cell types, although some are restricted to specific lineages.

Detailed analyses categorised immediate-early mitogen response genes into various groups according to their varying kinetics of expression. Genes such as c-*fos* (Cochran et al. 1984; Greenberg and Ziff 1984; Hogquist et al. 1991; Muller et al. 1984) and *egr*-1 (Cao et al. 1990; Christy et al. 1988; Seyfert et al. 1990; Sukhatme et al. 1987; Waters et al. 1990) are typically induced very rapidly and transiently. Their mRNAs appear within 15 minutes of mitogenic simulation and seldom persist for more than 1 hour. Both c-Fos and Egr-1 proteins are expressed over a similarly brief interval. Other immediate-early genes, such as c-*myc*, are induced somewhat later than c-*fos* and *egr*-1 (Kelly et al. 1983; Muller et al. 1984) but their expression persists for longer. In the case of c-*myc*, expression is maintained, albeit at a reduced level, throughout cell growth (Dean et al. 1986; Waters et al. 1991) (see below). Yet other genes comprise the so-called delayed-early group: their response requires *de novo* protein synthesis and their expression is substantially delayed within the G1 phase of the cell cycle.

The elegant and comprehensive analysis, characterisation and sequencing of the repertoire of immediate-early genes during fibroblast mitogenesis performed by Bravo and colleagues (Almendral et al. 1988) has demonstrated that immediate-early genes encode a diverse set of proteins. Various IE genes encode cytokines, cytokine receptors and signalling components, cytoskeletal proteins and a broad range of transcription factors. Presumably, many or all of these provide functions that are required for a quiescent cell to leave the quiescent state, enter the cell cycle and, where appropriate in the soma, communicate that fact to neighbouring cells in order to recruit them and establish correct tissue morphology.

Immediate-early genes that encode transcription factors are of particular interest because they constitute the key pleiotropic switching mechanisms that allow cells to alter their genetic programs and so provide long term adaptive changes to environmental and cytokine-mediated signals. Some of the known immediate-early mitogen-induced genes that encode transcription factors are listed in Table 1.

Action of immediate-early genes in mitogenesis

The identification of many proto-oncogenes as immediate-early mitogen response genes stands as one of the more important consolidating observations in recent cell biology. On the face of it, the mechanistic rationale seems clear: immediate-early mitogen response genes are integral in driving cells out of quiescence and

Table 1: Transcription factors encoded by immediate-early mitogen response genes

Archetype	Known Family Members	DNA Recognition site	DNA Binding Domain
fos	c-*fos*	TGA$(^{C}/_{G})$TCA (AP1)	bZ
	fos-B	"	bZ
	Fra-1	"	bZ
	Fra-2	"	bZ
jun	c-*jun*	TGA$(^{C}/_{G})$TCA (AP1)	
		"	
		"	bZ
myc	c-*myc*	CACGTG	bHLHZ
	N-*myc*	"	bHLHZ
	L-*myc*	"	bHLHZ
myb	c-*myb*	C$(^{C}/_{A})$GTT$(^{A}/_{G})$	Trp repeat
rel		GGG$(^{A}/_{G})$NT$(^{T}/_{C})(^{T}/_{C})$C	unique
	c-*rel*	"	unique
SRF	SRF	CC$(^{A}/_{T})_{6}$GG	MADS box
	RSRF	"	MADS box
ets	ets-1	GC$(^{C}/_{A})$GGAAGT	unique
	ets-2	"	unique
	erg	"	unique
	elk	$(^{C}/_{A})(^{C}/_{A})$GG$(^{A}/_{T})$	unique
	TCF	$(^{C}/_{A})(^{C}/_{A})$GG$(^{A}/_{T})$	unique
egr-1	egr-1, zif268, NGFIA	GCG$(^{G}/_{T})$GGGCG	C_2H_2 zinc finger

Key: *bZip - basic-leucine zipper* (Hurst 1994)
   bHLHZ- basic-Helix-Loop-Helix-leucine zipper (Littlewood and Evan 1994)

into proliferation. Accordingly, their inappropriate expression might be expected to promote the unrestrained proliferation of cells that typifies tumour cells. As a consequence, many IE genes can act as oncogenes. In particular, it can be inferred that those IE genes that encode transcription factors with oncogenic potential (e.g. c-*fos*, c-*jun*, c-*myc*, c-*rel*, c-*myb* and c-*ets*) have as their presumed genetic targets key genes controlling and orchestrating the processes of cell division and replication.

Whilst there is little doubt that this thesis is broadly correct, there remain a number of unresolved issues that complicate its interpretation. First, although the consensus DNA binding sites recognised by each IE transcription factor have been identified, such sequences are too short to provide sufficient specificity to allow direct identification of the repertoire of target genes controlled by each factor. Without elucidation of the panoply of genes regulated by a given transcription factor, and precisely how such target genes are regulated and in what context, it is not really possible to understand the *biological* role of that factor. At present, identification of target genes is piecemeal: the promoter and enhancer elements of any new gene can be dissected and sequenced to determine if they contain potential consensus binding sequences for any given transcription factor. Moreover, specific consensus binding sites often overlap with other known recognition elements, underscoring the importance of cross-talk and context in transcription factor activity. In addition, many consensus binding elements are shared by different transcriptional factors with differing biological roles. How specificity of action of such iso-specific transcription factors is controlled remains entirely unknown.

Second, mere induction of most IE genes appears insufficient for cell proliferation. Quiescent fibroblasts or lymphocytes triggered with mitogens for brief periods exhibit normal transient induction of the full panoply of immediate-early genes, yet do not become committed to traversing the cell cycle until some two hours before the onset of S phase (Pardee 1989). To reach this late G1 commitment point, often 12-24 hours after the classical immediate-early response, requires the sustained presence of a variety of mitogens and so-called "progression factors" whose natures differ from cell type to cell type. This introduces the notion that IE gene induction can be "futile" in certain circumstances - that is, unplugged from any downstream commitment.

Third, not only is induction of IE genes not sufficient for cell proliferation but in many cases IE gene expression does not even appear to be *necessary* for cell proliferation. For example, the ubiquitously induced mitogen response transcription factor c-Fos has been functionally inactivated both by introduction of dominant negative inhibitory mutants that block effective dimerisation and/or DNA binding of c-Fos to DNA (Ransone et al. 1990) and by gene ablation (Okada et al. 1994) and in neither case is cell proliferation necessarily inhibited. In general, it appears that many IE genes encode functions that are to a large degree redundant within individual cells. The significance of such redundancy will be further considered below.

Fourth, many IE genes are induced in cells by a wide range of stimuli including physical perturbation (Sadoshima et al. 1992) and insult (Andrews et al. 1987), irradiation (Anderson and Woloschak 1992; Hallahan et al. 1991; Marita et al.

1988), induction of differentiation (Mitchell et al. 1985), induction of programmed cell death (Kyprianou et al. 1991; Kyprianou et al. 1990), various neurological stimuli (Rusak et al. 1990) (Hunt et al. 1987) and many other signals (Cheung et al. 1989; Nambi et al. 1989). Few of these stimuli are associated with mitogenesis in any way. Moreover, many IE genes are induced in exponentially proliferating cells throughout the cell cycle by a variety of mitogenic and non-mitogenic cytokines. Thus, their induction is not restricted to the G0/G1 transition but seems to comprise part of a generic signal transduction pathway capable of responding to many different types of signal and with many different outcomes. This conclusion is reinforced by the fact that many IE genes are induced in different cells in response to stimuli that trigger non-proliferative responses (see below).

In summary, therefore, many immediate-early genes are induced by a variety of mitogenic and non-mitogenic signals in different cells types.

Immediate-early genes in neurons

One of the most telling indications that immediate-early growth response genes are not involved solely in regulating cell cycle entry arose from examination of the genetic events that ensued stimulation of nerve cells. Several initial studies demonstrated induction of c-*fos* in neurons within the CNS in response to a range of physiological and noxious stimuli. For example, c-*fos* is induced in a subset of spinal dorsal horn neurons following sensory nerve activation (Hunt et al. 1987). Such induction of c-*fos* is specific and is not observed in primary sensory or motor neurons implying that mere depolarisation is not sufficient for its induction. Moreover, over time, the pattern of c-*fos* immunoreactivity changes and tracks into the deeper laminae of the spinal cord and up into the brain (Williams et al. 1990a; Williams et al. 1990b). Other studies showed that c-*fos* is induced in CNS in response to convulsive drugs or strong electrical stimulation (Morgan et al. 1987; Morgan and Curran 1989). Because neurons in the adult CNS are incapable of proliferation, expression of c-*fos* cannot presage cell cycle entry of affected cells. Instead, c-*fos* expression appears to indicate some kind of long term adaptive or habituative response of neurons to stimuli, presumably related to the establishment of such recondite epiphenomema as memory. More recent studies have shown that many other immediate-early genes encoding transcription factors are induced by various stimuli in differing parts of the CNS (Wisden et al. 1990). Indeed, expression of immediate-early genes within neurons following specific stimuli can provide objective indicators of induction of various kinds of protracted neuronal activity. For example, the induction of c-*fos* induced in the dorsal horn neurons of the spinal cord by acute noxious stimuli can be largely abolished by pre-treatment of animals with high doses of morphiates that suppress pain (Tolle et al. 1994). Patterns of IE gene expression may therefore be useful both in determining neuronal circuitry and in assessing the efficacy of pharmacological agents.

In summary, the induction of a wide range of IE transcription factors in neurons indicates that these proteins are not solely involved in regulating cell cycle progression. Rather, they appear to function as generic "third messengers" that transduce a variety of different signals into a range of long term genetic changes in different cell types. The precise nature of these long term changes depends both upon the type of stimulus and upon the cell type on which the stimulus acts.

Redundancy in IE gene function and its significance

One of the most troubling and intriguing aspects of IE transcription factors is the apparent redundancy in their functions. This redundancy has been most graphically demonstrated by recent gene knockout experiments in which the ubiquitous IE transcription factors c-*fos* and c-*jun* have been deleted from the mouse genome (Hilberg et al. 1993; Okada et al. 1994). In the case of the c-*fos* knockout, resulting animals are surprisingly normal, and their somatic cells exhibit apparently normal proliferative control. In the case of the c-*jun* knockout, embryos die in mid-gestation. Nonetheless, virtually all somatic development occurs normally: the major exception being impaired liver development which is responsible for the observed embryonic death. The apparent dispensability of c-*fos* and c-*jun* seems in conflict with the observation that both of these genes are expressed in virtually all cell types in response to various extracellular stimuli; a fact that would seem to attest to their essential functions.

One part of the solution to this conundrum lies in the fact that multicellular organisms comprise an enormous diversity of differentiated cell types, each responsive to its own cadre of extracellular cues and signals and each fulfilling specific functions. Functions that appear redundant in specific cell types such as fibroblasts *in vitro* may not be similarly redundant in other cell lineages. This explanation is certainly consistent with the c-*jun* knockout data described above which indicates that although c-*jun* functionality is redundant in many tissues, it is certainly not so in developing liver.

Another component of redundancy is reflected by the fact that most immediate-early transcription factors are members of families which share many common features such as DNA sequence specificity and transcriptional activation (Table 1) (Hurst 1994; Littlewood and Evan 1994). In many cases, even though each member may possess its own subset of discrete properties, enough commonalty of function remains to permit differing members of each family to complement certain functions of other members. Arguably, however, a more important exposition of redundancy lies in an appreciation of the evolutionary pressures that drive or maintain redundant functions. A simplistic explanation for redundant functionality is to invoke the concept that redundancy arises to provide essential "backups" in case one system is disabled mutagenically. In this scenario, functions accrue redundancy if they are essential for viability and their loss would consequently be lethal. Whilst this paradigm for the mechanism of redundancy is persuasive and clearly applicable to certain situations, it is difficult to see it as an appropriate explanation for, say, redundancy amongst immediate-early genes in the process of mitogenesis. Acquisition of a lesion in the mitogenic machinery of any individual somatic mammalian cell is likely to have negligible consequences for the whole organism - there will always remain large numbers of normal cells present in any particular tissue that can replicate when necessary. As such, there is little or no evolutionary pressure to maintain redundant mitogenic machinery within somatic cells. In contrast, growth-restraining functions are likely to be highly redundant in somatic cells because loss of such a function in any single cell could result in neoplastic expansion of an individual clone and death for the whole organism. From such well-rehearsed arguments (Gould 1993) it seems most plausible that biological redundancy spontaneously arises under selectively neutral conditions as a consequence of the behaviour of DNA over evolutionary time (i.e. its tendency to duplicate itself). In organisms, such as vertebrates, in which genome size is not critical, there is effectively no selective pressure to remove duplicated functions. Such selective neutrality towards genetic duplication is most graphically demonstrated in the genus *Xenopus* in which it can extend to the entire genome (Kobel and Du Pasquier 1986). In summary, therefore, it appears that there is effectively no selective pressure against accumulating redundant or functionless DNA in vertebrates. However, once duplicated, individual genes then tend to evolve and append novel functions, albeit whilst retaining substantial functional overlap.

The general picture that emerges from consideration of the class of immediate-early genes exemplified by c-*fos* is of a transiently expressed generic third messenger that is widely expressed yet provides a function that in most cell types is neither essential nor solely responsible for many of the processes in which it takes part.

c-Myc - a paradigm for sustained immediate-early response

Involvement of c-*myc* in the G0/G1 transition from quiescence to proliferation was first suggested by its rapid induction (Kelly et al. 1983) together with other immediate-early growth-response genes, upon mitogenic stimulation of quiescent fibroblasts and lymphocytes. As already alluded to above, however, the pattern of c-*myc* expression following mitogenic stimulation differs from that of immediate-early growth response genes such as c-*fos* and *egr*-1 in that it is not restricted to a brief period at the G0/G1 transition (Fig. 2). Rather, it is continuously expressed in proliferating cells in a cell-cycle-independent manner (Dean et al. 1986; Rabbitts et al. 1985; Waters et al. 1991), suggesting a role in promoting continuous cell proliferation. Ectopic expression of c-*myc* drives quiescent fibroblasts into cycle without concomitant expression of other immediate-early growth response genes (Eilers et al. 1989; Eilers et al. 1991) and keeps them in cycle (Eilers et al. 1991; Evan et al. 1992). However, inhibition of c-*myc* expression in mitogen-stimulated lymphocytes with antisense oligonucleotide does not prevent G0/G1 transition events but blocks progression through G1 (Heikkila et al. 1987). Similarly, 3T3 fibroblasts transfected with the Tyr 809- mutant of CSF-1 receptor show normal ligand-stimulated tyrosine kinase activity and induction of c-*fos* and *jun* B, but no induction of c-*myc*. Such cells also exhibit a G1 block that can be relieved by ectopic c-*myc* expression (Roussel et al. 1991). Both of these experiments

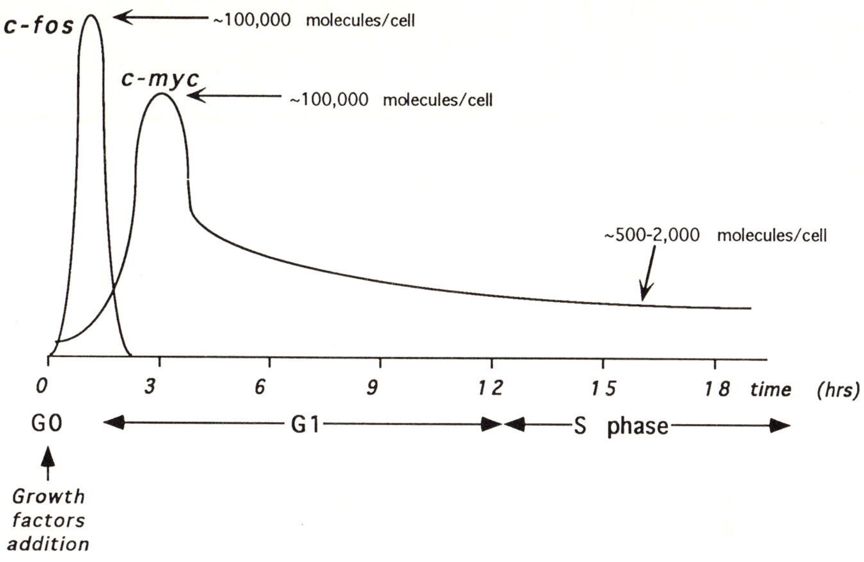

Fig. 2. Contrasting kinetics of induction and expression of c-*fos* and c-*myc* following sustained mitogenic stimulation of fibroblasts. Actual levels of each protein product in mouse human MRC-5 embryonic lung fibroblasts are shown.

imply a requirement for c-*myc* in G1 progression. From all these studies, certain general conclusions are evident. First, c-*myc* expression is essential for quiescent cells to proliferate. Second, c-*myc* expression appears not required for the G0/G1 transition but later in, and perhaps throughout, the cell cycle. Third, c-*myc* expression alone is sufficient to drive cell proliferation and by-pass, or leapfrog, the normal G0/G1 immediate-early growth response machinery.

c-*myc* expression is essential for growth of exponentially dividing cells. Steady-state c-*myc* expression in proliferating fibroblasts is continuously dependent upon mitogenic stimulation (Dean et al. 1986; Waters et al. 1991) and withdrawal of mitogens leads to the rapid and synchronous disappearance of c-*myc* mRNA and protein. This down-regulation of c-*myc* is not dependent of cell cycle position. Down-regulation of c-*myc* expression therefore comprises part of an immediate-early mitogen withdrawal response. Similarly rapid and cell cycle-independent down-regulation of c-*myc* occurs in response to anti-proliferative cytokines such as interferon-γ or TGFβ (Pietenpol et al. 1990).

c-*myc* encodes a short-lived sequence-specific DNA-binding transcription factor, c-Myc. c-Myc possesses an N-terminal transcriptional activation domain and a C-terminal DNA-binding/dimerisation bHLH-LZ domain similar to those present in several known transcription factors (Amati et al. 1994; Amati et al. 1992; Amati et al. 1992; Kato et al. 1990; Kretzner et al. 1992). However, the repertoire of

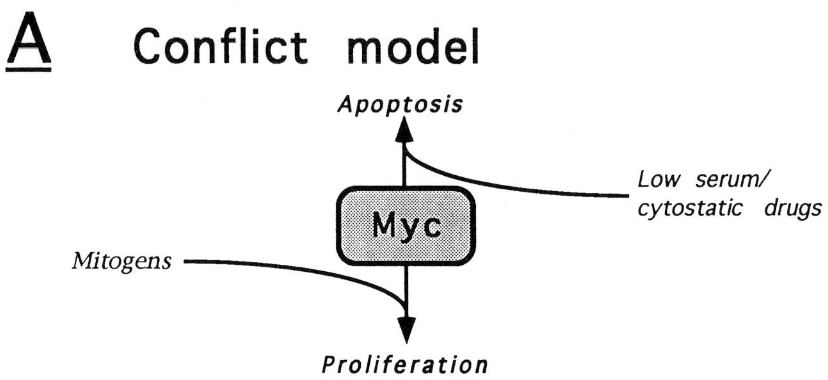

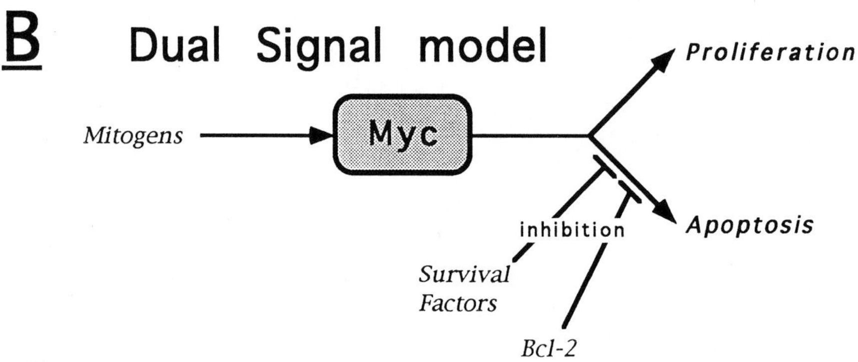

Fig. 4. Alternative models to explain induction of apoptosis by c-Myc.

levels of serum cytokines present in plasma or somatic tissue fluid are extremely low and never in excess. The attractive biological rationale for the "Dual Signal" model is that it intrinsically suppresses neoplastic transformation. Any mitogenic lesion in a dominant oncogene necessarily drives both cell growth *and* cell suicide. Thus, the affected clone will spontaneously delete itself when it outgrows the paracrine environment supplying it with survival factors - in effect, surveillance against neoplastic transformation is hardwired into the proliferative machinery.

Evidence favouring the "Dual Signal" model has recently been provided. Fibroblasts expressing c-Myc undergo apoptosis in the presence of cycloheximide or actinomycin D (Evan et al. 1992), implying that all necessary machinery for apoptosis pre-exists in cells as a consequence of c-Myc expression. Thus, the transcriptional apoptotic programme implemented by c-Myc does not arise as a consequence of a conflict in growth signals but is already present, although suppressed, in cells that exhibit no overt sign of apoptosis - for example fibroblasts expressing high levels of c-Myc growing in high serum. In turn, this implies that serum contains agents that suppress the apoptotic programme and we have recently identified these as insulin-like growth factors (IGFs) and PDGF (Harrington et al. 1994a). The abilities of IGFs and PDGF to block c-Myc-induced apoptosis are not dependent upon either factor's mitogenic activity and appear to represent a discrete signalling pathway that modulates cell viability independently of cell proliferation. The idea that cytokines that inhibit apoptosis do so in a manner unlinked to their mitogenic activity fits well with the known survival-potentiating activities of both IGF-1 and PDGF in post-mitotic cells such as neurons (Barres et al. 1992; Raff et al. 1993). We presume that other cell lineages are dependent upon different survival factors, for example, interleukin-3 appears to be a major anti-apoptotic cytokine for many haematopoietic lineages (Collins et al. 1992; Ormerod et al. 1992; Rodriguez et al. 1992).

Other types of immediate-early response

The cellular decisions to arrest growth, differentiate and undergo programmed cell death all appear to have genetic components to them which exhibit themselves as immediate-early responses. Induction of differentiation and growth arrest in haematopoietic cells triggers expression of several unique immediate-early response genes. Some of these genes, such as Myd-118 which is induced during myeloid differentiation (Abdollahi et al. 1991; Lord et al. 1990a; Lord et al. 1990b; Lord et al. 1990c; Lord et al. 1990d), are related to genes induced by growth arrest signals, and by DNA damage (Fornace et al. 1989) through a p53-dependent mechanism (Kastan et al. 1992). Others, such as Myd116 (Lord et al. 1990d), are related to immediate-early genes expressed by certain viruses (Chou and Roizman. 1992). Yet other genes, such as Mcl-1 and A1 which are also rapidly and transiently induced by differentiation factors (Kozopas et al. 1993; Lin et al. 1993), probably regulate cell viability as they are relatives of the anti-apoptotic proto-oncogene *bcl*-2 (Harrington et al. 1994b). The induction of cell death has been linked with expression of both c-*fos* (Smeyne et al. 1993) and c-*myc* (Amati et al. 1994; Askew et al. 1991; Bennett et al. 1993; Evan et al. 1992; Harrington et al. 1994b) proto-oncogenes and the cytotoxic cytokine TNF induces its own unique cadre of immediate-early genes (Sarma et al. 1992). Immediate-early responses therefore accompany the whole range of biological options available to cells.

References

Abdollahi A, Lord KA, Hoffman-Liebermann B, Liebermann DA (1991) Sequence and expression of a cDNA encoding MyD118: a novel myeloid differentiation primary response gene induced by multiple cytokines. Oncogene 6:165-167

Almendral JM, Sommer D, MacDonald-Bravo H, Burckhardt J, Perera J, Bravo R (1988) Complexity of the early genetic response to growth factors in mouse fibroblasts. Mol Cell Biol 8:2140-2148

Amati B, Brooks M, Levy N, Littlewood T, Evan G, Land H (1993) Oncogenic activity of the c-Myc protein requires dimerisation with Max. Cell 72:233-245

Amati B, Dalton S, Brooks M, Littlewood T, Evan G, Land H (1992) Transcriptional activation by c-Myc oncoprotein in yeast requires interaction with Max. Nature 359:423-426

Amati B, Littlewood T, Evan G, Land H (1994) The c-Myc protein induces cell cycle progression and apoptosis through dimerisation with Max. EMBO J 12:5083-5087

Anderson A, Woloschak GE (1992) Cellular proto-oncogene expression following exposure of mice to gamma rays. Radiat Res 130:340-4

Andrews GK, Harding MA, Calvet JP, E.D. A (1987) The heat shock response in HeLa cells is accompanied by elevated expression of the c-*fos* proto-oncogene. Mol Cell Biol 7:452-3458

Askew D, Ashmun R, Simmons B, Cleveland J (1991) Constitutive c-*myc* expression in IL-3-dependent myeloid cell line suppresses cycle arrest and accelerates apoptosis. Oncogene 6:1915-1922

Barres BA, Hart IK, Coles HS, Burne JF, Voyvodic JT, Richardson WD, Raff MC (1992) Cell death in the oligodendrocyte lineage. J Neurobiol 23:1221-30

Bennett M, Evan G, Newby A (1993) Deregulated expression of the c-*myc* oncogene abolishes inhibition of proliferation of rat vascular smooth muscle cells by serum reduction, interferon-g, heparin and cyclic nucleotide analogues and induces apoptosis. Circulation 74:525-536

Cao K, Koski RA, Gashler A, McKiernan M, Morri. CF, Gaffney R, Hay RV, Sukhatme VP (1990) Identification and characterization of the Egr-1 gene product, a DNA-binding zinc finger protein induced by differentiation and growth signals. Mol Cell Biol 10:1931-1939

Cheung HS, Mitchell PG, Pledger WJ (1989) Induction of expression of c-*fos* and c-*myc* protooncogenes by basic calcium phosphate crystal: effect of beta-interferon. Cancer Res 49:134-138

Chou J, Roizman B (1992) The gamma 1(34.5) gene of herpes simplex virus 1 precludes neuroblastoma cells from triggering total shutoff of protein synthesis characteristic of programed cell death in neuronal cells. Proc Natl Acad Sci USA 89:3266-3270

Christy BA, Lau LF, Nathans D (1988) A gene activated in mouse 3T3 cells by serum growth factors encodes a protein with 'zinc finger' sequences. Proc Natl Acad Sci USA 85:7857-7861

Cochran BH, Zullo J, Verma IM, Stiles CD. (1984) Expression of the c-*fos* gene and of an *fos*-related gene is stimulated by platelet-derived growth factor. Science 226:1080-1082

Cohen J, Duke R, Fadok V, KS S (1992) Apoptosis and programmed cell death in immunity. Ann Rev Immunol 10:267-293

Collins MK, Marvel J, Malde P, Lopez-Rivas A (1992) Interleukin 3 protects murine bone marrow cells from apoptosis induced by DNA damaging agents. J Exp Med 176:1043-1051

Coppola JA, Cole MD (1986) Constitutive c-*myc* oncogene expression blocks mouse erythroleukaemia cell differentiation but not commitment. Nature 320:760-763

Dean M, Levine RA, Ran W, Kindy MS, Sonenshein GE, Campisi J (1986) Regulation of c-*myc* transcription and mRNA abundance by serum growth factors and cell contact. J Biol Chem 261:9161-6

Denis N, Blanc S, Leibovitch MP, Nicolaiew N, Dautry F, Raymondjean M, Kruh J, Kitzis A (1987) c-*myc* oncogene expression inhibits the initiation of myogenic differentiation. Exp Cell Res 172:212-217

Dmitrovsky E, Kuehl WM, Hollis GF, Kirsch IR, Bender TP, Segal S (1986) Expression of a transfected human c-*myc* oncogene inhibits differentiation of a mouse erythroleukaemia cell line. Nature 322:748-750

Eilers M, Picard D, Yamamoto KR, Bishop MJ (1989) Chimaeras of Myc oncoprotein and steroid receptors cause hormone-dependent transformation of cells. Nature 340:66-68

Eilers M, Schirm S, Bishop JM (1991) The Myc protein activates transcription of the alpha-prothymosin gene. EMBO J 10:133-141

Evan G, Littlewood T (1993) The role of c-*myc* in cell growth. Curr Opin Genet & Dev 3:44-49

Evan G, Wyllie A, Gilbert C, Littlewood T, Land H, Brooks M, Waters C, Penn L, Hancock D (1992) Induction of apoptosis in fibroblasts by c-*myc* protein. Cell 63:119-125

Fornace JAJ, Nebert DW, Hollander MC, Luethy JD, Papathanasiou M, Fargnoli J, Holbrook NJ (1989) Mammalian genes coordinately regulated by growth arrest signals and DNA-damaging agents. Mol Cell Biol 9:4196-4203

Freytag SO (1988) Enforced expression of the c-*myc* oncogene inhibits cell differentiation by precluding entry into a distinct predifferentiation state in G0/G1. Mol Cell Biol 8:1614-1624

Freytag SO, Dang CV, Lee WMF (1990) Definition of the activities and properties of c-*myc* required to inhibit cell differentiation. Cell Growth & Diff 1:339-343

Gould SJ (1993) Eight little piggies (Penguin Science Series). Penguin, London

Greenberg ME, Ziff EB (1984) Stimulation of 3T3 cells induces transcription of the c-*fos* proto-oncogene. Nature 311:433-438

Hallahan DE, Sukhatme VP, Sherman ML, Virudachalam S, Kufe D, Weichselbaum RR (1991) Protein kinase C mediates X-ray inducibility of nuclear signal transducers Egr1 and Jun. Proc Natl Acad Sci USA 88:2156-2160

Harrington E, Fanidi A, Bennett M, Evan G (1994a) Modulation of Myc-induced apoptosis by specific cytokines. EMBO J 13:3286-3295

Harrington E, Fanidi A, Evan, G (1994b) Oncogenes and cell death. Curr Opin Genet Dev 4:120-129

Heikkila R, Schwab G, Wickstrom E, Loke SL, Pluznik DH, Watt R, Neckers LM (1987) A c-*myc* antisense oligodeoxynucleotide inhibits entry into S phase but not progress from G0 to G1. Nature 328:445-449

Hilberg F, Aguzzi A, Howells N, Wagner EF (1993) c-*jun* is essential for normal mouse development and hepatogenesis. Nature 365:179-181

Hogquist KA, Nett MA, Unanue ER, Chaplin DD (1991) Interleukin 1 is processed and released during apoptosis. Proc Natl Acad Sci USA 88:8485-8489

Hunt SP, Pini A, Evan GI (1987) Induction of c-*fos*-like protein in spinal cord neurons following sensory stimulation. Nature 328:632-634

Hurst HC (1994) Transcription factors 1: bZip proteins. Protein Profile. Ed. Sheterline P, Vol 1 Issue 2. Academic Press, London

Kastan MB, Zhan Q el DW, Carrier F, Jacks T, Walsh WV, Plunkett BS, Vogelstein B, Fornace AJ (1992) A mammalian cell cycle checkpoint pathway utilizing p53 and GADD45 is defective in ataxia-telangiectasia. Cell 71:587-597

Kato GJ, Barrett J, Villa GM, Dang CV (1990) An amino-terminal c-*myc* domain required for neoplastic transformation activates transcription. Mol Cell Biol 10:5914-5920

Kelly K, Cochran BH, Stiles CD, Leder P (1983) Cell specific regulation of the c-*myc* gene by lymphocyte mitogens and platelet-derived growth factor. Cell 35:603-610

Kobel H, Du Pasquier L (1986) Genetics of polyploid *Xenopus*. Trends Genet 2:310-315

Kozopas KM, Yang T, Buchan HL, Zhou P, Craig RW (1993) MCL1, a gene expressed in programmed myeloid cell differentiation, has sequence similarity to BCL2. Proc Natl Acad Sci USA 90.3516-3520

Kretzner L, Blackwood E, Eisenman R (1992) Myc and Max possess distinct transcriptional activities. Nature 359:426-429

Kyprianou N, English HF, Davidson NE, Isaacs JT (1991) Programmed cell death during regression of the MCF-7 human breast cancer following estrogen ablation. Cancer Res 51:162-166

Kyprianou N, English HF, Isaacs JT (1990) Programmed cell death during regression of PC-82 human prostate cancer following androgen ablation. Cancer Res 50.3748-3753

Lau LF, Nathans D (1985) Identification of a set of genes expressed during the G0/G1 transition of cultured mouse cells. EMBO J 4:3145-3151

Lin E, Orlofsky A, Berger M, Prystowsky M (1993) Characterization of A1, a novel hemopoietic-specific early-response gene with sequence similarity to *bcl*-2. J Immunol 151:1979-1988

Littlewood TD, Evan GI (1994) Transcription factors 2: helix-loop-helix proteins. Protein Profile. Ed. Sheterline P, Vol 1 Issue 6. Academic Press, London

Lord KA, Abdollahi A, Hoffman LB, Liebermann DA (1990a) Dissection of the immediate-early response of myeloid leukemia cells to terminal differentiation and growth inhibitory stimuli. Cell Growth Differ 1:637-645

Lord KA, Hoffman LB, Liebermann DA (1990b) Complexity of the immediate-early response of myeloid cells to terminal differentiation and growth arrest includes ICAM-1, Jun-B and histone variants. Oncogene 5:387-396

Lord KA, Hoffman-Liebermann B, Liebermann DA (1990c) Nucleotide sequence and expression of a cDNA encoding MyD88, a novel myeloid differentiation primary response gene induced by IL6. Oncogene 5.1095-1097

Lord KA, Hoffman-Liebermann B, Liebermann DA (1990d) Sequence of MyD116 cDNA: a novel myeloid differentiation primary response gene induced by IL6. Nucleic Acids Res 18:2823-2828

Marita B, Rahmsdorf HJ, Litfin M, Karin M, Herrlich P (1988) Activation of the c-*fos* gene by UV and phorbol ester: different signal transduction pathways converge to the same enhancer element. Oncogene 3:301-311

Mitchell RL, Zokas L, Schreiber RD, Verma IM (1985) Rapid induction of expression of proto-oncogene *fos* during human monocyte differentiation. Cell 40:209-217

Mohn KL, Laz TM, Hsu JC, Melby AE, Bravo R, Taub R (1991) The immediate-early growth response in regenerating liver and insulin-stimulated H-35 cells: comparison with serum-stimulated 3T3 cells and identification of 41 novel immediate-early genes. Mol Cell Biol 11:381-390

Morgan JI, Cohen DR, Hempstead JL, Curran T (1987) Mapping patterns of c-*fos* expression in the central nervous system after seizure. Science 237:192-197

Morgan JI, Curran T (1989) Stimulus-transcription coupling in neurons: role of cellular immediate-early genes. Trends Neurosci 12:459-462

Muller R, Bravo R, Burckhardt J, Curran T (1984) Induction of c-*fos* gene and protein by growth factors precedes activation of c-*myc*. Nature (London) 312:716-720

Nambi P, Watt R, Whitman M, Aiyar N, Moore JP, Evan GI, Crooke S (1989) Induction of c-*fos* protein by activation of vasopressin receptors in smooth muscle cells. Febs Lett 245:61-64

Okada S, Wang ZQ, Grigoriadis AE, Wagner EF, von Ruden T (1994) Mice lacking c-*fos* have normal hematopoietic stem cells but exhibit altered B-cell differentiation due to an impaired bone marrow environment. Mol Cell Biol 14:382-390

Ormerod MG, Collins MK, Rodriguez TG, Robertson D (1992) Apoptosis in interleukin-3-dependent haemopoietic cells. Quantification by two flow cytometric methods. J Immunol Methods 153:57-65

Pardee AB (1989) G1 events and regulation of cell proliferation. Science 246:603-608

Pietenpol JA, Stein RW, Moran E, Yaciuk P, Schlegel R, Lyons RM, Pittelkow MR, Munger K, Howley PM, Moses HL (1990) TGF-beta 1 inhibition of c-*myc* transcription and growth in keratinocytes is abrogated by viral transforming proteins with pRB binding domains. Cell 61:777-785

Rabbitts PH, Watson JV, Lamond A, Forster A, Stinson MA, Evan G, Fischer W, Atherton E, Sheppard R, Rabbitts TH (1985) Metabolism of c-*myc* gene products: c-*myc* mRNA and protein expression in the cell cycle. EMBO J 4:2009-2015

Raff M, Barres B, Burne J, Coles H, Ishizaki Y, Jacobson M (1993) Programmed cell death and the control of cell survival: lessons from the nervous system. Science 262:695-700

Ransone LJ, Visvader J, Wamsley P, Verma IM (1990) Trans-dominant negative mutants of Fos and Jun. Proc Natl Acad Sci USA 87:3806-3810

Rodriguez GT, Collins MK, Garcia I, Lopez RA (1992) Insulin-like growth factor-I inhibits apoptosis in IL-3-dependent hemopoietic cells. J Immunol 149:535-540

Roussel MF, Cleveland JL, Shurtleff SA, Sherr CJ (1991) Myc rescue of a mutant CSF-1 receptor impaired in mitogenic signalling. Nature 353:361-363

Rusak B, Robertson HA, Wisden W, Hunt SP (1990) Light pulse that shift rhythms induce gene expression in the suprachiasmatic nucleus. Science 248:1237-1240

Sadoshima J, Jahn L, Takahashi T, Kulik TJ, Izumo S (1992) Molecular characterization of the stretch-induced adaptation of cultured cardiac cells. An *in vitro* model of load-induced cardiac hypertrophy. J Biol Chem 267:10551-10560

Sarma V, Wolf FW, Marks RM, Shows TB, Dixit VM (1992) Cloning of a novel tumor necrosis factor-alpha-inducible primary response gene that is differentially expressed in development and capillary tube-like formation in vitro. J Immunol 148:3302-3312

Seyfert VL, McMahon S, Glenn W, Cao X, Sukhatme VP, Monroe JG (1990) *Egr-1* expression in surface Ig-mediated B cell activation. J Immunol 145:3647-3653

Smeyne R, Vendrell M, Hayward M, Baker S, Miao G, Schilling K, Robertson L, Curran T, Morgan J (1993) Continuous c-*fos* expression precedes programmed cell-death *in vivo*. Nature 363:166-169

Spencer CA, Groudine M (1991) Control of c-*myc* regulation in normal and neoplastic cells. Adv Cancer Res 56:1-48

Sukhatme VP, Kartha S, Toback FG, Taub R, Hoover RG, Tsai-Morris CH (1987) A novel early growth response gene rapidly induced by fibroblast, epithelial cell and lymphocyte mitogens. Oncogene Res 1:343-355

Tolle T, Schadrack J, Castro-Lopes J, Evan G, Roques B, Zieglgansberger W (1994) Effects of kelatorphan and morphine before and after noxious-stimulation on immediate-early gene-expression in rat spinal-cord neurons. Pain 56:103-112

Touchette N (1992) Dying cells reveal new role for cancer genes. J NIH Research 4:48-52

Waters C, Hancock D, Evan G (1990) Identification and characterisation of the *egr-1* gene product as an inducible, short-lived, nuclear phosphoprotein. Oncogene 5:669-674

Waters C, Littlewood T, Hancock D, Moore J, Evan G (1991) c-*myc* protein expression in untransformed fibroblasts. Oncogene 6:101-109

Williams S, Evan GI, Hunt SP (1990a) Spinal c-*fos* induction by sensory stimulation in neonatal rats. Neurosci Lett 9:309-314

Williams S, Evan GI, Hunt SP (1990b) Changing patterns of c-*fos* induction in spinal neurons following thermal cutaneous stimulation in the rat. Neuroscience 36:73-81

Wisden W, Errington M, Williams S, Dunnett S, Waters C, Hancock D, Evan G, Bliss T, Hunt SP (1990) Differential expression of immediate-early genes in the hippocampus and spinal cord. Neuron 4:603-614

Immediate-early gene activation as a window on mechanism in the nervous system

*S.P. Hunt, *L.A. McNaughton, R. Jenkins, and W. Wisden* [1]
Division of Neurobiology, Laboratory of Molecular Biology, MRC Centre, Hills Road, Cambridge CB2 2QH and *National Institute of Medical Research, Laboratory of Developmental Neurobiology, The Ridgeway, Mill Hill, London NW7 1AA, UK

Introduction

Immediate early genes (IEGs) were originally described as a class of genes rapidly and transiently expressed in cells stimulated with growth factors without the requirement for de novo protein synthesis (Cochran et al. 1983). c-Fos (Curran et al. 1987), c-Jun (Angel et al. 1988; Ryseck et al. 1988; Sakai et al. 1989) and other IEGs have been shown to be transcription factors (Chiu et al. 1988; Halzonetis et al. 1988; Sassone-Corsi et al. 1988; Abate et al. 1990; Benbrook and Jones 1990; Macgregor et al. 1990) and are differentially expressed in the central nervous system following specific types of stimulation (Hunt et al. 1987; Williams et al. 1989; Wisden et al. 1990; Cole et al. 1989, Tölle et al. 1994; Morgan and Curran 1992). Indeed, evidence is accumulating to suggest that IEG changes in gene expression within the nervous system signal long term adaptation within particular neural pathways.

Rapid and transient expression are obviously ideal characteristics for putative cellular 'activity markers' in that the pattern of expression could, because of the rapidity of expression, be assumed to be generated directly by the stimulus and not mediated polysynaptically or by some other process or be the residual trace of previous stimulation. However, while the value of IEGs as activity markers *in vitro* cannot be disputed, their appearance *in vivo* is under far greater control and appears to be tied to the physiological context of the stimulus. In this chapter we argue that the localization of IEG protein and mRNA in neurons and glial cells grown *in vitro* may accurately reflect that the cell has been recently stimulated in some way while the appearance of IEG product *in vivo* does not simply reflect a pattern of evoked activity but a pattern which is crucially related to the type of stimulation and the physiological state of the animal. In other words IEGs cannot be regarded simply as activity markers *in vivo* but perhaps as indicating the occurrence of a significant environmental stimulation that requires a long term change in certain aspects of neuronal physiology.

---

[1]

An invaluable contribution to the work presented here was made by our colleague Simon Williams who died tragically in 1992.

In vitro studies

An enormous number of studies have now been published on the *in vitro* use of IEGs (particularly c-Fos) to map activity at the single cell level in cell cultures and we do not intend to exhaustively review these here. Generally c-Fos immuno-cytochemistry coupled with localization or measurement of mRNA levels has been used to dissect out some aspect of cell physiology. In the examples given here c-Fos expression has been used to confirm the presence of excitatory amino acid (EAA) receptors on glial cells and neurons, allow the dissection of intracellular pathways to the nucleus and to look for the presence of trk growth factor receptors on neurons. It seems likely that this would not have been possible *in vivo* and indeed available evidence suggests that cells behave rather differently *in vivo* when stimulated in a comparable way.

For example, we have studied the effect of excitatory amino acids on the expression of mRNA for the immediate-early genes c-*fos*, c-*jun*, *jun* B, and NGFI-A (*zif*/268) in isolated type 1 cortical astrocytes (McNaughton and Hunt 1992). Excitatory amino acid receptors have been divided into a number of major subtypes; the metabotropic (quisqualic acid) receptor (QA, also known as GluR 1-6), the AMPA receptor and low affinity kainate receptor (GluR 1-4), and the high affinity kainic acid (KA) receptor (GluR 5-7, KA-1 KA-2) and N-methyl-D-aspartate (NMDA) receptor (Nakanishi 1992; Wisden and Seeburg 1993a,b; Young and Fagg 1990).

Astrocytes have been shown to increase their levels of intracellular Ca^{2+} in response to the stimulation by quisqualate, kainate and glutamate but not NMDA (Jensen and Chiu 1990; McNaughton et al. 1990). Further, electrophysiological studies have shown the presence of quisqualate, kainate but not NMDA receptors in cortical astrocytes *in vitro* and suggest that the response may be mediated by the activation of receptor-linked ion channels (reviewed by Barres 1991; Burnashev et al. 1992; Muller et al. 1992; Bowman and Kimelberg 1984; Sontheimer et al. 1988). In neuronal cell lines and neurons, elevation of Ca^{2+} levels is associated with the expression of genes such as c-*fos*. In PC12 cells, nicotinic or high K^{+} stimulation results in the secondary activation of voltage gated calcium channels which allow the entry of Ca^{2+} and a subsequent rise in intracellular Ca^{2+} levels (Greenberg et al. 1986; Bartel et al. 1989; Morgan and Curran 1986, 1991).

In previous studies we (McNaughton et al. 1990) and others (Jensen and Chiu 1990; Sontheimer et al. 1988: Wyllie et al. 1991) have been able to show that quisqualate and kainate, but not NMDA receptor stimulation of glial cells results in an elevation of intracellular calcium levels by two routes. Firstly, entry of calcium from the extracellular medium, presumably following opening of voltage sensitive calcium channels as in PC12 cells (Morgan and Curran 1986) and secondly by activation of the inositol phospholipid second messenger cascade, resulting in the generation of inositol triphosphate and the release of intracellular calcium from internal stores. We therefore examined the relationship between glutamate agonist stimulation and the expression of immediate-early genes.

The expression of the different genes was induced by 100 mM kainate, quisqualate, AMPA and high concentrations of K^{+} (140 mM). NMDA did not induce the expression of any of the genes studied. The effect of quisqualate stimulation was not inhibited by the antagonist CNQX or by withdrawal of external Ca^{2+} suggesting that activation was via metabotropic receptors. In contrast, the kainate effect was abolished by CNQX but not by the removal of external Ca^{2+} suggesting that the ionotropic receptor was in some way activating release from intracellular stores. However, elevated K^+ induced c-Fos only when calcium was present in the external medium, implying calcium movement through voltage dependent calcium channels. These findings also suggest that type-1 astrocytes lack NMDA receptors and that the induction of genes by quisqualate and kainate is complex, in part independent of the presence of calcium in the external medium and mediated through second messenger pathways following metabotropic or ionotropic receptor activation (Fig. 1).

We also show that c-*fos*, c-*jun* and *jun* B gene induction by QA and KA is independent of the presence of external Ca^{2+}. High K^+ concentrations also induced the expression of these genes, but unlike QA and KA, the presence of external Ca^{2+} was required.

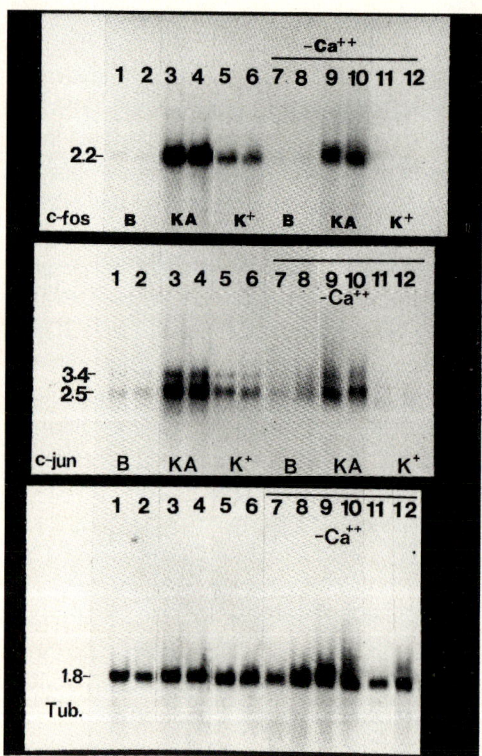

Fig. 1. Northern Blot analysis of the effect of extracellular calcium on gene expression. Lanes show the effect of kainic acid (100 mM) and potassium (140 mM) on c-*fos* and c-*jun* mRNA expression in the presence of external calcium (lanes 1-6) or in the absence of extracellular calcium (lanes 7-12). (From McNaughton and Hunt 1992)

Thus, while the expression of c-Fos and in fact a large number of IEGs can be predictably induced in glial cells *in vitro* and give significant insights into the intracellular mechanisms that accompany these receptor mediated events, this is entirely different from the *in vivo* state. *In vivo* immediate-early gene activation in glial cells is never seen in either early postnatal or adult animals following comparable stimulation paradigms (unpublished observations). Thus, what is seen in culture may be entirely different from the pattern of gene expression found in the brain. Nevertheless, these *in vitro* phenomena are extremely useful as markers of single cell activity.

IEG activation has been used to chart the second messenger pathways leading to the nucleus following glutamate or other types of stimulation of hippocampal neurons (Lerea and McNamara 1993; Bading et al. 1993). Primarily these studies demonstrated that the two distinct pathways of calcium entry into the neuron (via the NMDA receptor channel or through the opening of voltage dependent calcium channels (VSCCs) following membrane depolarization) resulted in the activation of different intracellular pathways which converge upon the c-*fos* gene. However, elevation of c-Fos expression occurs through two distinct regions of the c-*fos* promoter. Calcium entry through the NMDA receptor acts in part through a MAP kinase pathway (Bading et al. 1991) and induces changes in c-fos expression through the serum response element on the c-fos gene promoter, while calcium entry through VSCC appears to preferentially activate calcium-calmodulin dependent kinase II which influences c-Fos expression through the cyclic AMP response element on the c-*fos* gene. The assumption was that c-*fos* gene expression always follows EAA stimulation or high potassium treatment and appears to be born out by these studies. However in the brain the situation appears to be rather more complicated. The induction of long term potentiation in dentate gyrus granule cells following brief tetanic stimulation of the perforant pathway is dependent on NMDA activation and results in the induction of the IEG NGFI-A (also known as Egr-1 and Zif/268) but not for the most part c-Fos or c-Jun (Cole et al. 1989; Wisden et al. 1990; Worley et al. 1993). While it could be argued that adjusting the stimulation intensity would result in the expression of other IEGs, the remarkable fact is that LTP is only associated in this stimulation paradigm with the expression of NGFI-A (Fig. 2). It is highly unlikely that this observation could

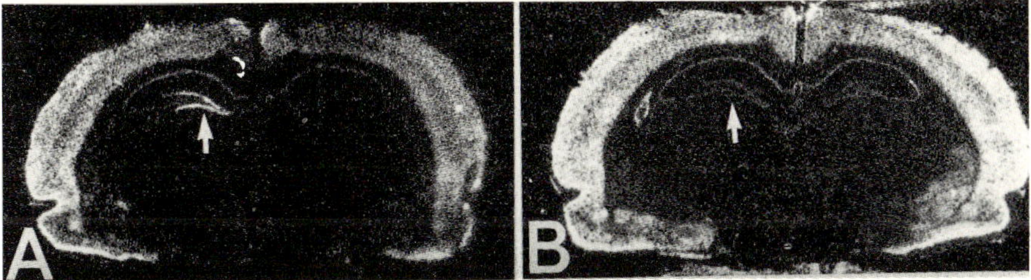

Fig. 2. (A) Induction of NGF1-A expression in the dentate gyrus of the hippocampus (arrow) following brief tetanic stimulation of the perforant pathway 30 min earlier. (B) Gene expression and LTP were blocked by prior treatment with AP5, an antagonist at the NMDA receptor. *In situ* hybridization was used to map the expression of NGF1-A mRNA. (From Wisden et al. 1990)

have been reproduced *in vitro* either in dissociated hippocampal neurons or brain slices. (In fact one of the major disappointments of c-Fos immunohistochemistry is that it has so far proved impossible to show gene expression in acute adult tissue slices maintained and stimulated *in vitro*, unpublished observations).

As a final example, Ip et al. (1993) have used c-Fos histochemistry and mRNA measurements to reinforce the evidence that the majority of hippocampal neurons *in vivo* and *in vitro* possess trkB and trkC but not trkA receptors. Hippocampal neurons grown *in vitro* were exposed to growth factors NGF, BDNF and NT3. c-Fos activation was only seen after addition of BDNF and NT3 corresponding to activation of the trkB and trkC receptors but not after addition of NGF because of the absence of trkA receptor. Note that the expression of c-Fos predictably followed the stimulation of the available receptor subtypes and in fact assumes this.

In vivo studies

Some years ago we were able to demonstrate that c-Fos is expressed post-synaptically in dorsal horn neurons of the spinal cord following noxious stimulation (Hunt et al. 1987; Williams et al. 1989, 1990a, 1990b). The protein product appears within 1-2 hours post-stimulation and c-Fos positive neurons are restricted to laminae I and II (the substantia gelatinosa) of the dorsal horn with some labelling in lamina V (Fig. 3). We found that the type of stimulation was crucial for a change in gene expression within postsynaptic neurons of the dorsal horn. Brief stimulation (5 sec) of high threshold C or Aδ sensory fibres but not low threshold A fibers results in substantial long term changes in spinal cord physiology. A similar brief noxious stimulus (chemical, heat or mechanical) results, within 1 hour, in the induction of c-Fos protein and of mRNA and protein for a large number of other IEGs within superficial and deep neurons of the ipsilateral dorsal horn. Induction of IEGs was never seen within sensory neurons in the dorsal root ganglion even following direct electrical stimulation. Similarly we have rarely seen c-Fos immunoreactivity within neurons in areas of termination of large diameter (low threshold) primary sensory fibres, including the dorsal column nuclei and the ventral horn. The pattern of c-Fos positive cells was established at birth (Fig. 4) and the number of labelled neurons was proportional to stimulus duration and intensity. These observations highlight the specificity of the gene induction within the spinal cord and suggest that it only occurs following stimulation that could be interpreted as injurious and was solely mediated by polymodal nociceptive C fibres and perhaps Aδ fibres.

This would not have been predictable from *in vitro* work. Moreover by 24h following noxious heat (but not chemical) stimulation or section of the peripheral nerve, this pattern of apparently monosynatically induced c-Fos in superficial neurons of the spinal cord gives way to a 'second wave' of labelled cells now restricted to deeper laminae and distributed bilaterally within the cord (Williams et al. 1990a) (Fig. 5).

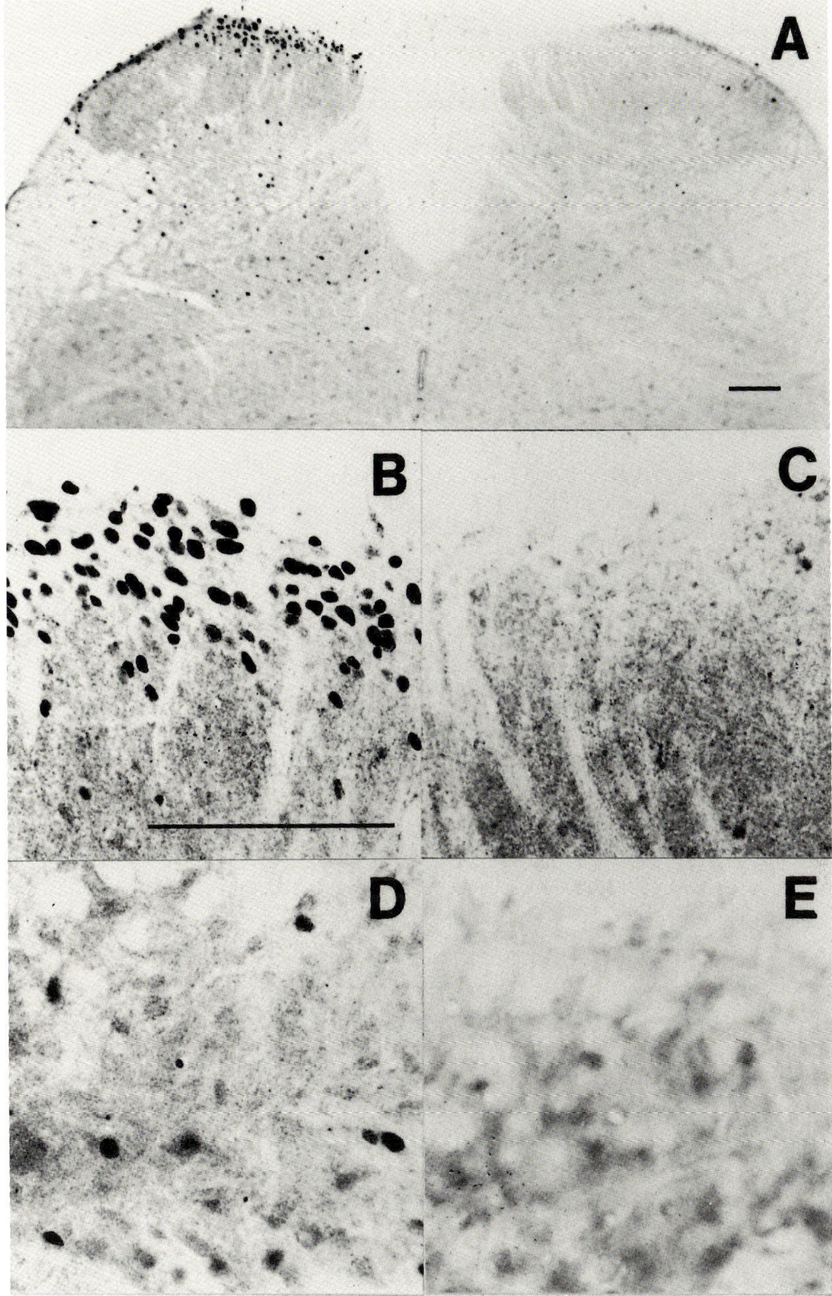

Fig. 3. The L_4 segment of the spinal cord 2 h after heat stimulation of the left paw of a rat under barbiturate anaesthesia. (A) Low power view showing Fos immunoreactivity concentrated ipsilaterally in superficial laminae, at higher power in (B), with a few positive cells in lamina V (D). No significant staining is seen contralaterally in superficial (C) or deep laminae (E). Both scale bars = 100 mm. (From S. Williams, unpublished)

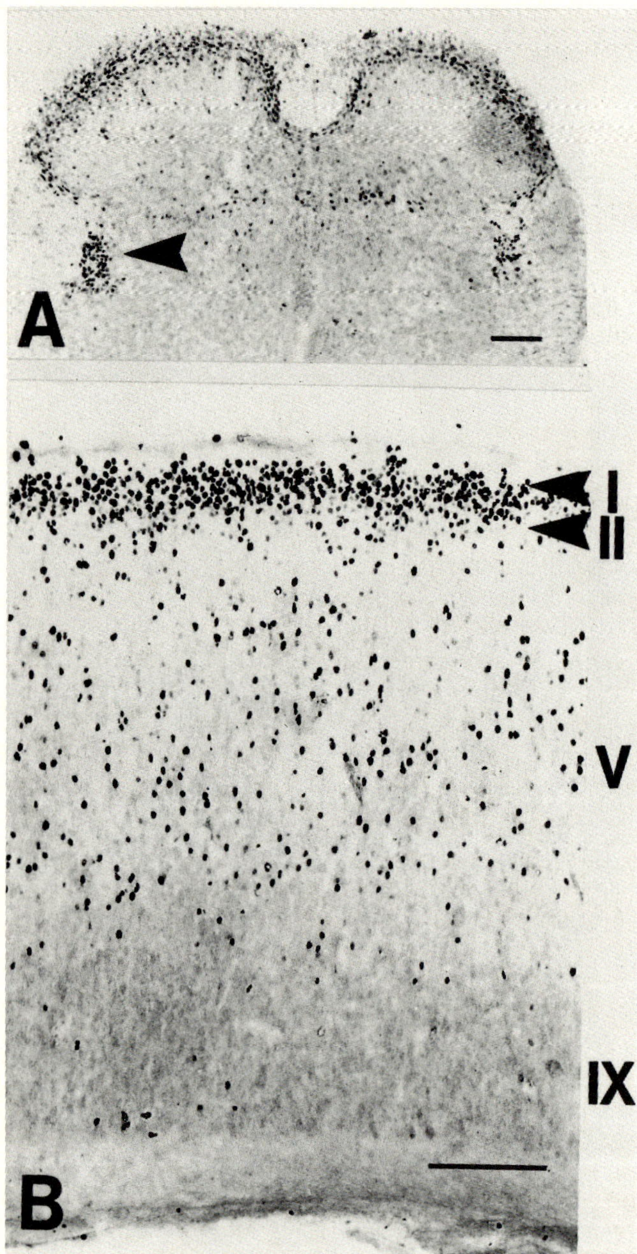

Fig. 4. (A) Section of rat caudal lumbar spinal cord at postnatal day 1 immunostained for c-Fos protein, 2h after subcutaneous injection of 50mg/kg capsaicin which stimulates the small diameter polymodal nociceptive sensory population. Nuclear staining is seen throughout laminae I and II and the arrow points to a cluster of c-Fos positive neurons in the lateral part of lamina V. Scale bar = 60 μm. (B) Parasagittal section at postnatal day 3, 2 h following subcutaneous plantar injection of 50 μl 4% formalin into ipsilateral hindpaw. Scale bar = 70 mm. (From Williams et al. 1991)

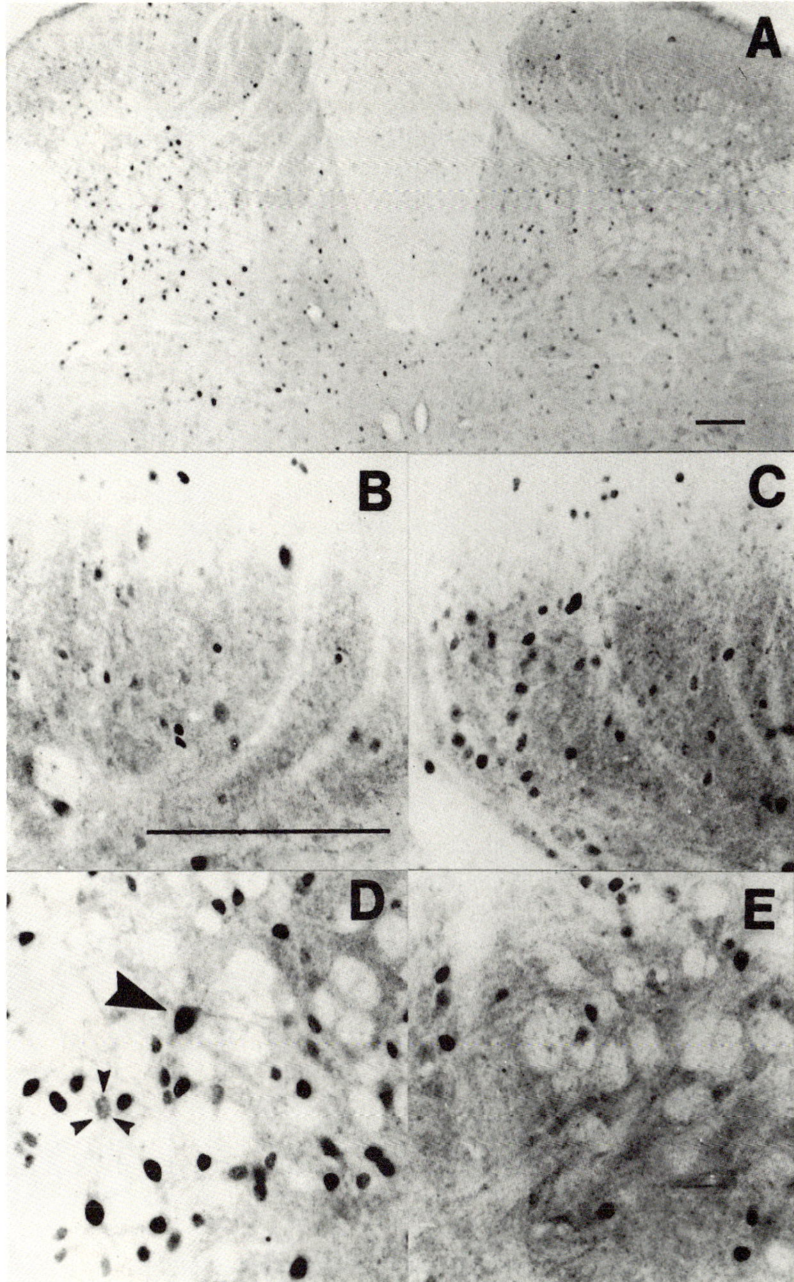

Fig. 5. 16 h after heat stimulation of the left hind paw of the rat. A) c-fos positive cells are now sparse in superficial laminae I-II (B, compare figure 2) but are present bilaterally within deeper lying neurons within laminae V-VII. (C) contralateral neurons within lamina II, and in lamina V of the stimulated (D) and unstimulated (E) sides of the cord. Scale bars = 100 mm.

This suggests that following a single noxious stimulation, a series of long term molecular changes evolve over time and involve neurons several synapses removed from the initial monosynaptic excitation. The initial pattern of c-*fos* gene expression is shared by a number of other IEG products including members of the Jun family (c-Jun, Jun B and Jun D), c-*fos* related genes (Fos Related Antigens (FRAs) and the zinc finger containing gene product NGFI-A (Zif/268) (Wisden et al. 1990). The kinetics of induction can vary between the different genes (Herdegen et al. 1991) and be differentially modified by morphine pretreatment (Tolle et al. 1994).

In animal models of chronic pain, such as that induced in rats by adjuvant injection, c-Fos positive neurons are similarly found within the deeper laminae of the dorsal horn and in amounts which parallel the severity of the 'disease' state (Abbadie and Besson 1992) suggesting that c-*fos* activation can be used as a marker for nociceptor activity in this model. However, while this observation might suggest that these deep lying, bilaterally distributed c-Fos positive neurons may contribute to the hyperalgesia and other clinical symptoms of the arthritic condition, such a conclusion may be premature. We have examined a second model involving the loose ligation of the sciatic nerve (Bennett and Xie 1988) which results in mechanical hyperalgesia around 6 days later and lasts for many weeks (Hunt et al. 1994). In fact, previous studies have indicated that within days of multiple ligation, there is abnormal electrical activity generated within the DRG (Kajander et al. 1992) and there is a substantial bilateral increase in metabolic activity within intermediate laminae of the spinal cord (Mao et al. 1992). However, at 28 days we were totally unable to find evidence for c-*fos* activation in deep spinal cord neurons. The use of c-Fos histochemistry therefore suggests that the two disease states have different molecular sequelae within the spinal cord which may indicate that different pharmacological routes have to be taken to control the pain associated with the two conditions.

Noxious stimulation always appears to result in c-*fos* activation within the spinal cord. However, in several other well studied systems the pattern of c-*fos* activation seen after stimulation is dependent upon the physiological status of the animal. Three examples of this will suffice.

The expression of c-Fos protein was examined in the basal forebrain of male rats 60 min following intracerebroventricular injection of angiotensin II. A distinct pattern of c-Fos expression was seen in the anterior region of the third ventricle, medial preoptic region, the subfornical region and a number of hypothalamic nuclei including the supraoptic nuclei, paraventricular nucleus and in the amygdala (Herbert et al. 1992). Allowing animals to drink during the 60 min survival period modified this pattern of response in that the c-Fos expression was reduced within the supraoptic nucleus and the paraventricular nucleus but not other labelled brain areas mentioned. That is the physiological state of the animal was paramount in determining the expression of c-Fos in certain areas of the brain.

Brennan et al. (1992) reported on the expression of c-Fos in the accessory olfactory bulb. Female mice form a memory for the pheromones of the male with which they mate. The site of synaptic change appears to be the accessory olfactory bulb. During the process of memory formation, transient increases in c-Fos expression were seen in this area. This increase in the number of cells required the association between pheromones *and* mating, neither stimulation alone being sufficient.

Finally, c-Fos is induced by light in the suprachiasmatic nucleus which is the site of an endogenous circadian pacemaker responsible for the timing and overt expression of daily rhythms (Rusak et al. 1990; Takeuchi et al. 1993). Light entrains or synchronizes mammalian circadian rhythms through a specialized retino-hypothalamic projection that terminates in the ventral portion of the nucleus (Kornhauser et al. 1992). A flash of light in the subjective night of a rat or hamster results in the appearance of c-Fos immunoreactivity within the suprachiasmatic nucleus (Fig. 6). However when night is extended for 12h into what is effectively subjective day and the light flash repeated, then there is no c-Fos expression (Rusak et al. 1990). Thus the ability of light to alter expression of c-Fos is regulated by the circadian clock. The mechanism by which this is accomplished as with the regulation of c-Fos during angiotensin II stimulation or c-Fos expression during mating is at present unknown but underscores the fundamental principle that IEG expression *in vivo* has to take into account the physiological state of the animal.

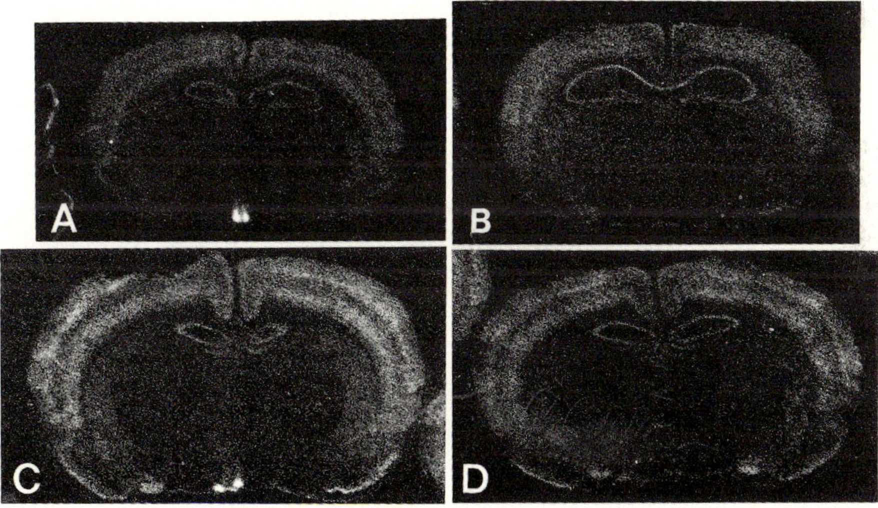

Fig. 6. c-fos expression in the suprachiasmatic nucleus of the brain of the hamster (A,B) and rat (C,D) in control animals (B, D) and in animals exposed to light during the middle of the subjective night (dark phase) (A, C). Note the massive expression of c-fos mRNA following stimulation. (From Rusak et al. 1990)

Long term expression of IEGs in the DRG

IEGs are a diverse group of genes grouped together because of their characteristic kinetics of expression and their independence from new protein synthesis. This mode of expression is an ideal characteristic for a gene that can in some cases be used to indicate monosynaptic changes in neurons. However, it has become clear that certain of these genes, particularly c-*jun*, can be expressed only after a considerable delay following stimulation and remain high for long periods of time, for example during the differentiation of F9 carcinoma cells with retinoic acid or in the neuronal response to axonal damage (Yang-Yen et al. 1990; Jenkins and Hunt 1991; Leah et al. 1991; Sleigh 1992). A number of recent studies have implicated the c-*jun* proto-oncogene in the control of growth and differentiation in a variety of cell types including neurons and glial cells (Bos et al. 1990; Castellazzi et al. 1991; De Felipe et al. 1993; De Felipe and Hunt 1994; De Groot et al. 1990; Jenkins and Hunt 1991; Jenkins et al. 1993a-c; Leah et al. 1991; Wong et al. 1992; Yamaguchi-Iwai et al. 1990). Following axon damage or block of axonal transport in peripheral sensory or motor neurons in the rat, there is a delayed expression of c-*jun* protein and mRNA which is maintained until the nerve has fully regenerated (Jenkins and Hunt 1991; Leah et al. 1991; Herdegen et al. 1993) (Fig. 7).

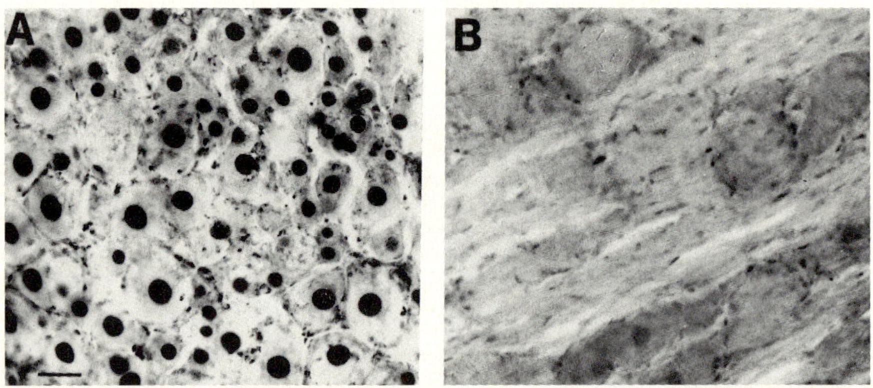

Fig. 7. c-Jun expression in the dorsal root ganglion L_4, 4 d following section of the sciatic nerve (A) and in contralateral control ganglion (B) in the rat. Scale bar = 25 mm. (From Jenkins et al. 1993c)

This is never accompanied by the expression of c-Fos. Damage to the sciatic nerve induces expression of c-*jun* protein and mRNA within DRG cells and motor neurons that give rise to that nerve (Jenkins and Hunt 1991; Leah et al. 1991; Herdegen et al. 1993). Similar expression of c-Jun was also seen in nigrostriatal and rubrospinal neurons following lesion of their axons (Jenkins et al. 1993a, b). General axonal transport block, induced by the application of colchicine to the sciatic nerve, resulted in the expression of c-Jun in both large and small diameter fibres, suggesting that the absence of a transported factor may play a role in the expression of c-Jun in sensory neurons. c-Jun expression was maintained if the axons were prevented from regenerating. However, the role of c-Jun in the maintenance of the regenerative response is unclear and we therefore designed a number of experiments to test the hypothesis that c-Jun expression is related to regeneration or growth of axons rather than simply axonal damage.

In a recent study (Jenkins et al. 1993c) we looked at a peripheral nerve lesion, which can be prevented from regenerating by ligation and thereby allow nearby intact sensory neurons to undergo collateral sprouting into denervated peripheral tissue. Previous work suggests that large axons (conducting in the Aβ range) do not undertake any significant sprouting, whereas some fine axons do. In this study we ligated the sciatic nerve (which projects through dorsal root ganglia L4, L5, L6), and looked for a reaction in the nearby intact saphenous nerve afferents (most of which are derived from the adjacent L3 dorsal root ganglion).

Sciatic nerve lesion at 24h but not 2h, resulted in a massive expression of c-Jun in the nuclei of all damaged sensory neurons as has previously been described in detail (Jenkins and Hunt 1991; Leah et al. 1991; Herdegen et al. 1993). Similar results were seen in ganglion L3 following saphenous nerve section. From 24 hours the expression of c-Jun was approximately constant until the time of reinnervation of target, at which point it declined to control levels. c-Jun protein was also expressed in Schwann cell nuclei of the sciatic nerve segment distal but not proximal to the nerve lesion (De Felipe and Hunt 1994).

It was also of interest to know whether the c-Jun response is correlated with axon damage alone, or with neuronal growth. Our demonstration that c-Jun expression is elevated in the L3 DRG after sciatic nerve section suggests the latter. There is no evidence of axons of the L3 DRG projecting through the sciatic nerve (Molander and Grant 1986), so changes in this ganglion are likely to reflect expression in undamaged neurons. There is considerable, if somewhat conflicting, evidence that some collateral sprouting of intact afferents can occur into denervated tissue. After sciatic nerve section and ligation, fine, but not coarse, axons in the saphenous nerve are reported to progressively invade some of the sciatic territory (Devor et al. 1979; Kinnman and Aldskogious 1986). The elevation of c-Jun in a limited number of cells, particularly small cells, in the L3 DRG, is therefore well-correlated with such sprouting.

The function of c-Jun in neuronal regeneration is unclear. We recently (Jenkins et al. 1993c) pointed out that the suggestion of a causal relationship between other changes in gene expression seen in sensory neurons following axotomy and c-Jun expression may be an oversimplification. Most of these changes in gene expression were restricted to subsets of sensory neurons and occurred at different times after axotomy. For example, the upregulation of GAP 43 expression occurs initially in small diameter sensory neurons and involves larger diameter DRGs only at longer survival times (Sommervaille et al. 1991). In contrast, the change in c-Jun expression was rapid and appeared simultaneously in both small and large sensory neurons within 24 hours, regardless of chemical phenotype (Jenkins et al. 1991; Leah et al. 1991). Thus it seems unlikely that c-Jun expression leads directly to changes in the expression of other downstream genes simply by direct interaction with the relevant promoter sequences.

In summary there is evidence that IEGs such as *c-jun* can be expressed and regulated with non IEG kinetics. In the case of *c-jun* there appears to be a close relationship with regeneration and the neuron and glial cell response to damage.

Conclusions

We have attempted to demonstrate that IEG expression in neurons and glial cells must be interpreted with care. *In vitro*, these gene products, particularly c-Fos, come close to fulfilling all the characteristics of activity markers. However *in vivo* such a simplistic analysis cannot be made and the physiological context of the animal must be taken into consideration. Nevertheless the modulation of a pattern of gene expression induced by particular forms of physiologically relevant stimulation provide unique insights into the way that information is being stored and processed within the central nervous system. There is also the problem that IEGs can operate with non-immediate early kinetics as in the case of c-*jun*. In this case the expression of the gene may be signalling the attempt of the cell to regenerate and survive and in fact this *in vivo* gene expression can be reproduced in adult cultures of dorsal root ganglion cells (De Felipe et al. 1994).

We have largely ignored the problem of what IEG expression could mean for individual neurons and glial cells. There is little direct evidence to work with. IEGs are transcription factors which bind to their relevant DNA sequence on the regulatory region of certain genes and influence gene expression. Recent studies also suggest that genes such as c-*fos* and c-*jun* can act as selectors of cell responsivness to external stimuli (Diamond et al. 1990) or may compete out more potent combinations of proteins to effectively repress gene expression (Kerppola et al. 1993). Most of these possibilities remain to be explored in the nervous system.

References

Abate C, Luk D, Gentz R, Rauscher FJ III, Curran T (1990) Expression and purification of the leucine zipper and DNA-binding domains of Fos and Jun: both Fos and Jun contact DNA directly. Proc Nat Acad Sci USA 87: 1032-1036

Abbadie C, Besson JM (1992) c-Fos expression in rat lumbar spinal cord during the development of adjuvant-induced arthritis. Neurosci 48(4): 985-993

Angel P, Hattori K, Smeal T, Karin M (1988) The *jun* proto-oncogene is positively autoregulated by its product, Jun/AP-1. Cell 55:875-885

Bading H, Greenberg ME (1991) Stimulation of protein tyrosine phosphorylation by NMDA receptor activation. Science 253:902-904

Bading H, Ginty DD, Greenberg ME (1993) Regulation of gene expression in hippocampal neurons by distinct calcium signalling pathways. Science 260: 181-186

Barres B (1991) Glial ion channels. Curr Opinions in Neurobiol 1:354-359

Bartel DP, Sheng M, Lau LF, Greenberg ME (1989) Growth factors and membrane depolarization activate distinct patterns of early response genes: dissociation of *fos* and *jun* induction. Genes and Develop 3:304-313

Benbrook DM, Jones NC (1990) Heterodimer formation between CREB and Jun proteins. Oncogene 5:295-302

Bennett GJ, Xie YK (1988) A peripheral neuropathy in rat that produces disorders of pain sensation like those seen in man. Pain 33:87-108

Bos TJ, Monteclaro FS, Mitsuobu F, Ball AR, Chang CHW, Nishimura T, Vogt PK (1990) Efficient transformation of chicken embryo fibroblasts by c-Jun requires structural modification in coding and non-coding sequences. Genes and Develop 4:1677-1687

Bowman CL, Kimelberg HK (1984) Excitatory amino acids directly depolarize rat brain astrocytes in primary culture. Nature 311:656-659

Brennan PA, Hancock D, Keverne EB (1992) The expression of the immediate-early genes c-Fos, Egr-1 and c-Jun in the accessory olfactory bulb during the formation of an olfactory memory in mice. Neurosci 49: 277-284

Burnashev N, Khodorova A, Jonas P, Helm J, Wisden W, Monyer H, Seeburg PH, Sakmann B (1992) Calcium permeable AMPA kainate receptors in fusiform cerebellar glial cells. Science 256:1566-1570

Castellazzi M, Spyrou G, La Vista N, Dangy JP, Piu F, Yaniv M, Brun G (1991) Overexpression of c-*jun*, *jun* B or *jun* D. Proc Natl Acad Sci USA 88:8890-8894

Chiu R, Boyle WJ, Meek J, Smeal T, Hunter T, Karin M (1988) The c-Fos protein interacts with c-Jun/AP-1 to stimulate the transcription of AP-1 responsive genes. Cell 54:541-552

Cochran B H, Reffel AC, Stiles CD (1983) Molecular cloning of gene sequences regulated by platelet-derived growth factor. Cell 33:939-947

Cole AJ, Saffen DW, Baraban JM, Worley PF (1989) Rapid increase of an immediate-early gene messenger RNA in hippocampal neurons by synaptic NMDA receptor activation. Nature 340:474-476

Curran T, Gordon MB, Rubino KL, Sambucetti LC (1987) Isolation and characterization of the c-*fos* (rat) cDNA and analysis of post-translational modification in vitro. Oncogene 2:79-84

De Felipe C, Hunt SP (1994) The differential control of c-*jun* expression in regenerating sensory neurons and their associated glial cells. J Neurosci 114:2911-2923

De Felipe C, Jenkins R, O'Shea R, Williams TSC, Hunt SP (1993) The role of immediate-early genes in the regeneration of the central nervous system. Advances in Neurology, Raven Press, New York 263-271

De Groot RP, Kruyt FAE, van der Saag PT, Kruijer W (1990) Ectopic expression of c-*jun* leads to differentiation of P19 embryonal carcinoma cells. EMBO J 9(6):1831-1837

Devor M, Schonfield D, Seltzer Z, Wall PD (1979) Two modes of cutaneous reinnervation following cutaneous nerve injury. J Comp Neurol 185:211-220

Diamond MI, Miner JN, Yoshinaga SK, Yamamoto KR (1990) Transcription factor interactions: selectors of positive or negative regulation from a single DNA element. Science 249:1266-1271

Greenberg ME, Ziff EB, and Greene LE (1986) Stimulation of neuronal acetylcholine receptor induces rapid gene expression. Science 234:80-83

Halzonetis TD, Georgeopoulos K, Greenberg ME, Leder P (1988) c-jun dimerizes with itself and with c-fos forming complexes of different DNA binding abilities. Cell 55:917-924

Herbert J, Forskling ML, Howes SR, Stacey PM, Shiers HM (1992) Regional expression of c-Fos antigen in the basal forebrain following intraventricular infusions of angiotensin and its modulation by drinking either water or saline. Neurosci 51:867-882

Herdegen T, Brecht S, Mayer B, Leah J, Kummer W, Bravo R, Zimmermann M (1993) Long-lasting expression of Jun and Krox transcription factors and nitric oxide synthase in intrinsic neurons of the rat brain following axotomy. J Neurosci 13(10):4130-4145

Herdegen T, Tölle TR, Bravo R, Zieglgänsberger W (1991) Sequential expression of Jun B, Jun D and Fos B proteins in rat spinal neurons: a cascade of transcriptionl operations during nociception. Neurosci Lett 129:221-224

Hunt SP, Pini A, Evan G (1987) Induction of c-Fos like protein in spinal cord neurons following sensory stimulation. Nature 328:632-634

Hunt SP, Smith G, Bond A, Munglani R, Thomas K, Elliot P (1994) Changes in immediate-early genes, neuropeptide immunoreactivity and other neuronal markers in the spinal cord and dorsal root ganglion in a rat model of neuropathic pain. In: Hokfelt T, Schaible HG, Schmidt R (eds) Neuropeptides, Nociception and Pain. Elsevier, in press

Ip NY, Li Y, Yancopoulos GD, and Lindsay RM (1993) Cultured hippocampal neurons show responses to BDNF, NT-3, and NT-4, but not NGF. J Neurosci 13(8):3394-3405

Jenkins R, Hunt SP (1991) Long term increases in the levels of c-jun mRNA and Jun protein like immunoreactivity in motor and sensory neurons following axon damage. Neurosci Lett 129:107-111

Jenkins R, McMahon SB, Bond AB, Hunt SP (1993c) Expression of c-Jun as a response to dorsal root and peripheral nerve section in damaged and adjacent intact primary sensory neurons in the rat. Eur J Neurosci 5: 751-759

Jenkins R, O'Shea R, Thomas KL, Hunt SP (1993b) c-Jun expression of immediate-early genes in substantia nigra neurons following striatal 6-OHDA lesions in the rat. Neurosci 53:447-457

Jenkins R, Tetzlaff W, Hunt SP (1993a) Differential expression of immediate-early genes in rubrospinal neurons following axotomy in the rat. Eur J Neurosci 5:203-209

Jensen AM, Chiu SY (1990) Fluorescence measurement of changes in intracellular calcium induced by excitatory amino acids in cultured cortical astrocytes. J Neurosci 10:1165-1175

Kajander KC, Wakisaka S, Bennett GJ (1992) Spontaneous discharge originates in the dorsal root ganglion at the onset of a painful peripheral neuropathy in the rat. Neurosci Lett 138:225-228

Kerppola TK, Luk D, Curran T (1993) Fos is a preferential target of glucocorticoid receptor inhibition of AP-1 activity in vitro. Mol Cell Biol 13(6):3782-3791

Kinnmann E, Alskogius H (1986) Colateral sprouting of sensory axons in the glabrous skin of the hind paw after chronic nerve lesion in the adult and neonatal rat: a morphological study. Brain Res 377:73-82

Kornhauser JN, Mayo DE, Takahashi KE, J.S. (1992) Regulation of *jun*-B messenger RNA and AP-1 activity by light and a circadian clock. Science 255:1581-1584

Kouzarides T, Ziff E (1988) The role of the leucine zipper in the fos-jun interaction. Nature 336:646-651

Leah JD, Herdegen T, Bravo R (1991) Selective expression of Jun proteins following axotomy and axonal transport block in peripheral nerves in the rat; evidence for a role in the regeneration process. Brain Res 566:198-207

Lerea LS, McNamara JO (1993) Ionotropic glutamate receptor subtypes activate *c-fos* transcription by distinct calcium-requiring intracellular signaling pathways. Neuron 10:31-41

Macgregor PF, Abate C, Curran T (1990) Direct cloning of leucine zipper proteins: Jun binds cooperatively to the CRE with CRE-BP1. Oncogene 5: 451-458

Mao J, Price DD, Coghill RC, Mayer DJ, Hayes RL (1992) Spatial patterns of spinal cord [^{14}C]-2-deoxyglucose metabolic activity in a rat model of painful peripheral mononeuropathy. Pain 50:89-100

McGregor GP, Gibson SJ, Sabate IM, Blank MA, Christofedes ND, Wall PD, Polak JM, Bloom SR (1989) Effect of peripheral nerve section and nerve crush on spinal cord neuropeptides in the rat: increased VIP and PHI in the dorsal horn. Neurosci 13:207-216

McNaughton LA, Hunt SP (1992) Regulation of gene expression in astrocytes by excitatory amino acids. Mol Brain Res 16:261-266

Mollander C, Grant G (1986) Laminar distribution and somatotopic organization of primary afferent fibres from hind limb nerves in dorsal horn. A study by transneuronal transport of horseradish peroxidase. Neurosci. 19:297-312

Morgan JI, Curran T (1986) Role of ion flux in the control of c-*fos* expression. Nature 322:552-555

Morgan JI, Curran T (1991) Stimulus transcription coupling in the nervous system: involvement of the inducible proto oncogenes c-*fos* and c-*jun*. Ann Rev Neurobiol 14:421-451

Muller T, Moller T, Berger T, Schnitzer J, Kettenmann H (1992) Calcium channel entry through kainate receptors and resulting potassium channel blockade in Bergmann glial cells. Science 256:1563-1566

Nakanishi S (1992) Molecular diversity of glutamate receptors and implications for brain function. Science 258:597-603

Rusak B, Robertson H, Wisden W, Hunt SP (1990) Light pulses that shift rhythms induce gene expression in the suprachiasmatic nucleus. Science 248:1237-1240

Ryseck RP, Hirai SI, Yaniv M, Bravo R (1988) Transcriptional activation of c-*jun* during the G_0/G_1 transition in mouse fibroblasts. Nature 334:535-537

Sakai M, Okuda A, Hatayama I, Sato K, Nishi S, Muramatsu M (1989) Structure and expression of the rat c-*jun* messenger RNA: tissue distribution and increase during chemical hepatocarcinogenesis. Cancer Res 49:563-567

Sassone-Corsi P, R LJ, Lamph WW, Verma IM (1988) Direct interaction between Fos and Jun nuclear oncoproteins: role of the 'leucine zipper' domain. Nature 336:692-695

Sleigh MJ (1992) Differentiation and proliferation in mouse embryonal carcinoma cells. BioEssays 14:769-775

Sommervaille T, Reynolds ML, Woolf CJ (1991) Time-dependant differences in the increase in GAP-43 expression in dorsal root ganglion cells after peripheral axotomy. Neurosci 45:213-230

Sontheimer H, Kettenmann H, Backus KH, Schachner M (1988) Glutamate opens Na^+/K^+ channels in cultured astrocytes. Glia 1:328-336

Takeuchi JS, Aronin W, Schwartz NWJ (1993) Compositional changes of AP-1 DNA binding proteins are regulated by light in a mammalian circadian clock. Nature 11:825-836

Tölle TR, Herdegen T, Schadrack J, Bravo R, Zimmermann M, Zieglgänsberger W (1994) Application of morphine prior to noxious stimulation differentially modulates expression of Fos, Jun and Krox-24 proteins in rat spinal cord neurons. Neurosci 58:305-321

Williams TSC, Evan GI, Hunt SP (1990a) Changing patterns of c-fos induction in spinal neurons following thermal cutaneous stimulation in the rat. Neurosci 36:73-81

Williams TSC, Evan GI, Hunt SP (1990b) Spinal c-fos induction by sensory stimulation in neonatal rats. Neurosci Lett 109:309-314

Williams TSC, Evan GI, Hunt SP (1991) C-*fos* induction in the spinal cord after peripheral nerve lesion. Eur J Neurosci 3:887-894

Williams TSC, Pini A, Evan G, Hunt SP (1989) Molecular events in the spinal cord following sensory stimulation. Processing of sensory information in the superficial dorsal horn of the spinal cord. New York, Plenum Press. 273-284

Wisden W, Errington ML, Williams S, Dunnett SB, Waters C, Hitchcock D, Evan G, Bliss TV, Hunt SP (1990) Differential expression of immediate-early genes in the hippocampus and spinal cord. Neuron 4:603-614

Wisden W and Seeburg PH (1993a) Mammalian ionotropic glutamate receptors. Curr Opin Neurobiol 3:291-298

Wisden W and Seeburg PH (1993b) A complex mosaic of high-affinity kainate receptors in rat brain. J Neuroscience 13:3582-3598

Wong WY, Havarstein L, Morgan IM, Vogt PK (1992) c-Jun causes focus formation and anchorage-independant growth in culture but is non-tumorigenic. Oncogene 7:2077-2080

Worley PF, Bhat RV, Baraban JM, Erickson CA, McNaughton BL, Barnes CA (1993) Thresholds for synaptic activation of transcription factors in hippocampus: correlation with long term enhancement. J Neurosci 13: 4776-4786

Wyllie DJ, Mathie A, Symonds CJ, Cull-Candy SG (1991) Activation of glutamate receptors and glutamate uptake in identified macroglial cells in rat cerebellar cultures. J Physiol 432:235-258

Yamaguchi-Iwai Y, Satake M, Murakami Y, Sakai M, Muramatsu M, Ito Y (1990) Differentiation of F9 embryonal carcinoma cells induced by c-*jun* and activated c-Ha-*ras* oncogenes. Proc Natl Acad Sci USA 87:8670-8674

Yang-Ycn HF, Chiu R, Karin M (1990) Elevation of AP1 activity during F9 cell differentiation is due to increased c-*jun* transcription. The New Biologist 2(4):351-361

Young AB, Fagg GE (1990) Excitatory amino acid receptors in the brain: membrane binding and receptor autoradiographic approaches. Trends Pharmacol Sci 11:126-133

Differential expression of immediate-early genes during synaptic plasticity, seizures and brain injury suggests specific functions for these molecules in brain neurons

M. Dragunow
Department of Pharmacology, School of Medicine,
The University of Auckland, Auckland, NEW ZEALAND

Introduction

Although the discovery of immediate-early gene expression in brain and spinal cord neurons occured over 6 years ago (Dragunow et al. 1987; Ruppert and Wille 1987; Dragunow and Robertson 1987; Hunt et al. 1987; Morgan et al. 1987; White et al. 1987; Saffen et al. 1988), the functional role of the sub-group that generates DNA-binding proteins (the immediate-early gene proteins, IEGPs) in neurons remains obscure. We have taken a number of approaches to discovering the functions of these proteins including describing the physiological and pathological conditions in which different IEGPs are expressed, and the biochemical pathways (*e.g.*: neurotransmitters, receptors, second messengers, etc) involved in their expression in particular brain regions after specific inducing stimuli. The goal of this work is to then focus on the functional role of particular IEGPs in particular systems. Using drugs which block IEGP expression by interfering with the biochemical pathway through which they are induced we can initially test the role of a particular IEGP in a non-specific fashion. Subsequently, using antisense DNA technology *in vivo* to inactivate specific genes (Chiasson et al. 1992), we can directly test the role of a particular IEGP in our test system.

Because of the apparently widespread inducibility of the Fos IEGP (see Morgan and Curran 1991), many researchers have questioned the specificity and importance of this molecule to neuronal function. However, the DNA-binding activity of IEGPs is complex since they interact with many other transcription factors. Therefore, the induction of Fos, measured in isolation, by a variety of stimuli, is functionally meaningless since members of the Fos (c-fos, Fos B, FRA-1, FRA-2, FRA's) and Jun (c-Jun, Jun B, Jun D) family of transcription factors must form homodimers (Jun family only) and heterodimers (*e.g.*: Fos/Jun D) for transcriptional activity (Morgan and Curran 1991). Furthermore, a naturally occurring truncated form of Fos B can inhibit Fos/Jun transcriptional activity (Nakabeppu and Nathans 1991), increasing the possible interactions between these various transcription factors. Furthermore, Fos B, like Fos, can inhibit the c-fos promotor, but truncated Fos B cannot. Also, recent studies have found that Fos/Jun and Jun/Jun dimers bend DNA in opposite directions (Kerpolla and Curran 1991) and form topologically distinct DNA-protein complexes. Also, Fos and Jun act cooperatively with the glucocorticoid receptor but have opposite effects on transcriptional activity in combination with the glucocorticoid and mineralocorticoid receptors (Diamond et al. 1990; Funder et al. 1993). Fos and Jun also interact with other members of the ligand-dependent family of transcription factors such as the thyroid hormone receptor (Zhang et al. 1991). Furthermore, retinoic acid is a negative regulator of activator protein 1 (AP-1) responsive genes (Schule et al. 1991). Auwerx and Sassone-Corsi (1991) have recently demonstrated the existence of a phosphorylation-modulated inhibitor of Fos/Jun activity which they have called IP-1. Regulation by methylation has also been reported for the Egr-1 IEG (Seyfert et al. 1990), and this gene is also regulated by protein phosphorylation and dephosphorylation (Cao et al. 1992). Also, phosphorylation of c-Jun protein by mitogen-activated protein-serine (MAP) kinases positively regulates its transacting activity (Pulverer et al. 1991), whereas phosphorylation of c-Jun by casein kinase II inhibits Jun activity (Hunter and Karin 1992). Another level at which c-Jun can be regulated has recently been described by Baichwal et al. (1991) who demonstrated the existence of a cell-specific inhibitor of c-Jun activity that was overcome by SRC and RAS proteins. A factor called Ref-1 stimulates AP-1 DNA-binding activity of Fos and Jun without altering their DNA-binding specificity via a novel reduction-oxidation mechanism (Xanthoudakis and Curran 1992). Jun can also form dimers with Ca^{++}/cAMP response element binding protein (CREB) and activating transcription factors (ATF's, Hai and Curran 1991). Fos can also interact with a helix-loop-helix zipper protein (Blanar and Rutter 1992). Thus, the complexity of these interactions suggests that the IEGPs could potentially mediate a vast array of biological processes in neurons.

Therefore, to begin to understand the functions of these molecules we must determine the expression of a range of IEGPs in neurons and uncover the biochemical pathways of their induction. However, the specificity and function of IEGPs may not only be determined by the pattern of IEGPs induced, but also by their temporal profile of induction (especially for Fos and Jun family members that form dimers for transcriptional activity), and by the biochemical pathway through which they are induced. For example, the induction of the c-myc IEG in the presence of growth factors leads to cell proliferation, whereas c-myc induction in growth arrested cells induces apoptosis (programmed cell death, Bissonnette et al. 1992; Evan et al. 1992; Fanidi et al. 1992; Shi et al. 1992). Presumably, the biochemical environment in which a particular IEGP is induced will influence its DNA-binding activity and its biological effect. For this reason multiple biochemical induction pathways in neurons may not be solely a redundancy mechanism (Lerea et al. 1992), but may confer specificity and a wide range of functions to IEGPs. For example, Krox-24 is induced in CA1 pyramidal cells by tonic activation of NMDA receptors (Richardson et al. 1992) and by phasic activation of muscarinic receptors (Hughes and Dragunow 1993). Since the biochemical environment into which Krox-24 is induced following NMDA receptor activation will be different to that following muscarinic receptor activation (e.g.: cellular pH, ions, protein kinase activation, other transcription factors), the biological effect may be quantitatively and/or qualitatively different. Furthermore, because the microenvironment of neurons in different brain regions differs (growth factors, neurotransmitters, transcription factors, etc) IEGPs may have different biological effects in different brain regions. Thus, expression of c-Jun in dentate granule cells after long-term potentiation may be involved in LTP stabilization (Demmer et al. 1993), whereas c-Jun induction in CA1 neurons after hypoxia-ischaemia and status epilepticus may be involved in cell death (Dragunow et al. 1993, see later).

Our studies have focused upon the potential roles of IEGPs in synaptic plasticity and memory formation, seizure development, tardive dyskinesia, and nerve cell death. We have discovered that there is a differential induction of IEGPs following these various types of stimulation. Notably, LTP is associated with the induction of the Krox family of IEGPs (Krox-20 and Krox-24), whereas nerve cell death is associated with the induction of c-Jun. In contrast, activation of muscarinic receptors and block of dopamine receptors induces Jun B, but not c-Jun (Dragunow et al., 1993).

Synaptic plasticity - expression of Krox-20 and Krox-24

We have investigated the long-term potentiation (LTP) model of synaptic plasticity to try to uncover the role of IEGPs in this process.

Methods

The methods of surgery and LTP induction have been previously described (Douglas et al. 1988; Dragunow et al. 1989a, b; Jeffery et al. 1990; Abraham et al. 1991; Richardson et al. 1992; Demmer et al. 1993) and will not be described here.

Results

(see Table)
These studies were initially conducted in collaboration with Harry Robertson and Rob Douglas at Dalhousie University (Douglas et al. 1988). We showed that LTP in anaesthetized rats did not induce Fos/FRA's in dentate granule cells (Douglas et al. 1988) but that only bursting of granule cells was sufficient for Fos/FRA expression. Granule cells do not burst in response to perforant-path tetanization because of strong GABA-mediated recurrent inhibition. However, we showed that if this inhibition was removed with bicuculline then Fos/FRAs were induced in these neurons (Douglas et al. 1988). We extended these studies in New Zealand in collaboration with Cliff Abraham and Warren Tate of Otago University. We found that FRA's, c-jun, jun B, and jun D mRNA and protein could be induced after LTP in an NMDA receptor-dependent fashion, but only in unanaesthetized rats (Dragunow et al. 1989; Jeffery et al. 1990; Demmer et al. 1983). However, the induction of these IEG's correlated poorly with both the induction and mainte-

Temporal profile of IEGP induction

IEGPs show different time-courses of expression and decay after various treatments with Fos, Jun B and Krox-24 being induced first (within 1 hr) and falling away rapidly to baseline (within 4 hrs). Induction of Fos B and Jun D is more delayed (4 hrs) but these IEGPs remain elevated for longer (up to 24 hrs, Demmer et al. 1993; Dragunow et al. 1992; MacGibbon et al., 1994; Dragunow et al., 1993). This suggests that Fos/Jun B and Fos B/Jun D dimers are formed in neurons. Krox-20 is induced rapidly after LTP but levels remain elevated for over 24 hrs (Williams et al., 1994). c-Jun induction is delayed (3-6 hrs) and prolonged (24-72 hours) after status epilepticus and hypoxia-ischaemia. After axotomy, c-Jun induction is more delayed (48 hrs) and very prolonged (weeks - months depending upon brain region, Dragunow 1992; Leah et al., 1993).

Non-nerve cell expression of IEGPs

We discovered a number of years ago that mechanical injury to the brain induces Fos in non-nerve cells in white matter tracts (corpus callosum, fornix-fimbria, internal capsule) and grey matter (hippocampus, cortex, striatum, thalamus), and also in the lateral and third ventricles in ependymal cells and on the pial surfaces of the brain (Dragunow and Robertson 1988). Similar effects are observed after hypoxia-ischaemia (Gunn et al. 1990), and heat-shock (Dragunow et al. 1989c). More recently, we have discovered that this induction is differential with Fos/FRAs, c-Jun, and Krox-24 being strongly induced, Jun B and Jun D weakly induced and Fos B and Krox-20 not induced in non-nerve cells (Dragunow 1990; Dragunow and Hughes, 1993; Dragunow et al., 1993). Although induction in neurons after mechanical brain injury is dependent on NMDA receptor activation and Calcium/calmodulin kinase II, induction in non-nerve cells is not (Dragunow et al. 1990b). Cyclic AMP appears to be an important signal for IEGP induction in non-nerve cells in the brain (Dragunow and Faull 1989). Furthermore, using double-labelling methods we have discovered that most of the non-nerve cells expressing IEGPs are not GFAP-positive astrocytes or GPDH-positive oligodendrocytes (Dragunow et al. 1990c). Thus, at present the identity of the non-nerve cells expressing IEGPs in white and grey matter is unclear. However, a number of these cells appear from their doughnut-shaped morphology to be dividing (Dragunow and Hughes, 1993; Dragunow et al., 1993). Thus, IEGP expression in non-nerve cells after traumatic stimuli may be involved in the proliferation of these cells in response to injury as has been demonstrated in other cell types (Kovary and Bravo 1991; Riabowol et al. 1988).

Table of results: effects of various treatments on IEGP expression

TREATMENT	RECEPTOR	IEGP							
		FOS	FRA	FOSB	JUN	JUN B	JUN D	KROX-20	KROX-24
LTP	NMDA	0	+	0	+	++	+	+++	+++
HYPOXIA/ISCHAEMIA	NMDA/?	++	++	++	+++	+	+	?	+/-
BRIEF SEIZURES	?	+++	+++	+	+++	+++	+++	+++	+++
ST. EPILEPTICUS	?	++	++	++	+++	+	+	+	++
MK-801/PILOCARPINE-S.	?	++	++	++	+++	+	+	NM	++
AXOTOMY	?	0	0	0	+++	0	0	0	0
HALOPERIDOL	D2	+++	+++	++	0	+++	+	+	++
PILOCARPINE	M1	+++	+++	NM	0	+++	O	NM	++
MK-801	M1	+++	+++	NM	+	+++	NM	NM	+++
NON-NERVE CELLS	?	+++	+++	0	+++	+	+	0	+++

Key:

NM	= not measured.
-	= reduction below basal levels.
0	= no induction.
+	= slight induction.
++	= moderate induction.
+++	= strong induction.

Target genes

A number of potential target genes have been described (e.g.: tyrosine hydroxy-lase, TRH, CCK, GAD, proenkephalin, NGF, White and Gall 1987; Hengerer et al. 1990; Moochetti et al. 1989; Sonnenberg et al. 1989; Gall and Isaacson 1989; Rosen et al. 1992; Olenik et al. 1991; Najlirahim et al. 1991; Goc et al. 1992; Weiser et al. 1993). Although the proenkephalin gene has been proposed as an IEGP target (late-response gene, Sonnenberg et al. 1989) we have recently found that although hippocampal injury induces IEGPs in dentate granule cells, proenke-phalin mRNA is not induced (Hughes and Dragunow, unpublished observations).

After seizures, induction of IEGPs in dentate granule cells (Dragunow et al. 1992) may induce axonal sprouting (Sutula et al. 1992) by regulating the expression of structural and/or microtubule proteins, and/or growth factor production (Ernfors et al. 1991; Gall and Isaacson 1989). Expression of Jun D in somato-statin neurons after seizures (Dragunow et al. 1992) may be involved in regulation of the somatostatin gene.

In other studies we have found that BDNF mRNA is induced in many situations where the Fos, Jun and Krox families of IEGPs are induced with the BDNF mRNA induction being delayed with respect to the IEGP expression: following MK-801 injection (Hughes et al. 1993b); following LTP via NMDA receptor activation (Dragunow et al., 1993); and following hippocampal injury via NMDA receptors (Hughes et al., 1993c). These results suggest that BDNF may be a target

44

of IEGP expression. However, we (Hughes et al., 1993c) and others (Demello et al. 1992) have found that BDNF expression after hippocampal injury and seizures is not blocked by cycloheximide, a protein synthesis inhibitor which prevents IEGP induction (Hughes et al., 1993c), suggesting that BDNF is not an IEGP target, but is itself an IEG. As an IEG it appears to have both an interneuronal signalling and DNA binding capacity (Wetmore et al. 1991). Thus, BDNF may also function as a transcription factor in neurons. Interestingly, expression of BDNF after LTP correlates with LTP maintenance (Dragunow et al., 1993) and BDNF appears to be involved in spatial memory processes (Falkenberg et al. 1992).

Overall discussion and conclusion

Our studies over the past 7 years have described the physiological and pathological stimuli that induce members of the Fos (Fos, Fos B, FRAs), Jun (c-Jun, Jun B, Jun D) and Krox (Krox-20, Krox-24) families of IEGPs in neurons and non-nerve brain cells. We have described their anatomical and temporal profile of expression, and have uncovered neurotransmitters (dopamine, acetylcholine, glutamate) and receptors (D1, D2, M1, NMDA) that mediate their expression after various stimuli (LTP, seizures, brain injury).

Perhaps the most important result of these studies is the realization that different IEGPs are expressed differentially. This is particularly true for the Jun family of IEGPs where c-Jun is selectively expressed in dying neurons after hypoxia-ischaemia, status epilepticus, and axotomy, Jun D is selectively expressed in somatostatin-phenotype neurons after seizures, and Jun B is selectively expressed after activation of muscarinic receptors or block of dopamine receptors. Similarly, the Krox family (Krox-20 and Krox-24) are linked closely to LTP whereas members of the Jun and Fos families are not. These results suggest that specific IEGPs serve specific physiological (Krox and BDNF in memory formation) and pathological (Jun in nerve cell death; Jun D in seizure development; Jun B in tardive dyskinesia) functions in neurons. However, IEGPs can also be induced in the same neurons in a coordinate fashion as occurs after brief hippocampal seizures (Dragunow et al. 1992).

A second important discovery is that there are multiple induction pathways for IEGPs in neurons. In the cortex and limbic system acetylcholine and glutamate play major roles in the regulation of IEGP expression. Perhaps it is no coincidence that these neurotransmitters also subserve vital functions in learning and memory formation, seizure mechanisms and nerve cell death. In the basal ganglia, dopamine is an important regulator of IEGP expression, and these molecules may serve important functions associated with this important neurotransmitter (e.g.: movement control, psychosis, brain reward). Thus, IEGPs may function as critical components of the signal transduction cascade in neurons linking cell membrane receptors to changes in neuronal gene expression.

A third major focus of our work has been the demonstration that IEGPs are also differentially induced in non-nerve cells in the brain some of which appear to be in various stages of cell division. Although we are yet unsure of the identity of these cells, their expression of IEGPs after brain injury caused by hypoxia-ischaemia, status epilepticus, and mechanical trauma, and evidence of their proliferation suggests that IEGP expression in these cells may be important for cell proliferation after brain injury. They may also play a role in neuroprotective and/or neurotoxic growth factor/cytokine production by non-nerve cells after injury to the brain.

In conclusion, the past 7 years of IEGP research has generated a wealth of data suggesting that these molecules have specific functions in neurons. The challenge will be to demonstrate a causal link between IEGP expression and biological effect and to detect target genes regulated by these molecules.

Acknowledgements: The research described in this review was supported by grants from the NZ Health Research Council, the Auckland Medical Research Foundation, the New Zealand Neurological Foundation, and the Lottery Board - Health. I would like to thank the following collaborators, students and technicians who contributed to specific areas of the research work reviewed and described here:- Harry Robertson and Rob Douglas (Dalhousie University, Halifax, Canada); Cliff Abraham, Warren Tate, Betty Mason, and Dave Bilkey (Otago University, Dunedin, NZ); John Leah (Griffith University, Brisbane, Australia); Rodrigo Bravo (Princeton, USA); Peter Gluckman, Erica Beilharz, Chris Williams and Ernest Sirimanne (Auckland University, Auckland, NZ); and Trish Lawlor, Paul Hughes, Geraldine MacGibbon, Debbie Young and Kerin Singleton (Auckland University, Auckland, NZ).

References

Abraham WC, Dragunow M, Tate WP (1991) The role of immediate-early genes in the stabilization of long-term potentiation. Molecular Neurobiol 5:297-314

Abraham WC, Mason SE, Demmer J, Williams JM, Richardson CL, Tate WP, Lawlor PA, Dragunow M, (1993) Correlations between immediate-early gene induction and the persistence of long-term potentiation. Neuroscience 56:717-727

Auwerx J, Sassone-Corsi P (1991) IP-1: A dominant inhibitor of fos/jun whose activity is modulated by phosphorylation. Cell 64:983-993

Baichwal VR, Park A, Tjian R (1991) v-Src and EJ Ras alleviate repression of c-Jun by a cell-specific inhibitor. Nature 352:165-168

Bissonnette RP, Echeverri F, Mahboubi A, Green DR (1992) Apoptotic cell death induced by c-*myc* is inhibited by *bcl*-2. Nature 359:552-553

Blanar MA, Rutter WJ (1992) Interaction cloning: identification of a helix-loop-helix zipper protein that interacts with c-fos. Science 256:1014-1017

Cao X, Mahjendran R, Guy GR, Tan YH (1992) Protein phosphatase inhibitors induce the sustained expression of the Egr-1 gene and the hyperphosphorylation of its gene product. J Biol Chem 267:12991-12997

Chiasson BJ, Hooper ML, Murphy PR, Robertson HA (1992) Antisense oligonucleotide eliminates in vivo expression of c-fos in mammalian brain. Eur J Pharmacol 227:451-453

Choi DW (1990) Cerebral hypoxia: some new approaches and unanswered questions. J Neurosci 10, 8:2493-2451

Demello SR, Jiang C, Lamberti C, Martin SC, Heinrich G (1992) Differential regulation of the nerve growth factor and brain-derived neurotrophic factor genes in L929 mouse fibroblasts. J Neurosci Res 33 4:419-526

Demmer J, Dragunow M, Lawlor PA, Mason SE, Leah JD, Abraham WC, Tate WP (1992) Differential expression of immediate early genes after hippocampal long-term potentiation in awake rats. Mol Brain Res 17:279-286

Dessi F, Charriaut-Marlangue C, Ben-Ari Y (1992) Anisomycin and cycloheximide protect cerebellar neurons in culture from anoxia. Brain Res 581:323-326

Diamond MI, Miner JN, Yoshinaga SK, Yamamoto KR (1990) Transcription factor interactions: selectors of positive or negative regulation from a single DNA element. Science 249:1266-1271

Douglas RM, Dragunow M, Robertson HA (1988) High-frequency discharge of dentate granule cells, but not long-term potentiation, induces c-fos protein. Mol Brain Res 4:259-262

Dragunow M, Williams M, Faull RLM. (1990d) Haloperidol induces fos and related molecules in intrastriatal grafts derived from fetal striatal primordia. Brain Res 530:309-311

Dragunow M, Abraham WC, Goulding M, Mason SE, Robertson HA, Faull RLM (1989a) Long-term potentiation and the induction of c-fos mRNA and proteins in the dentate gyrus of unanesthetized rats. Neurosci Lett 101:274-280

Dragunow M, Robertson HA (1987a) Kindling stimulation induces c-fos protein(s) in granule cells of the rat dentate gyrus. Nature 329:441-442

Dragunow M, Robertson HA (1987b) Generalized seizures induce c-fos protein(s) in mammalian neurons. Neurosci Lett 82:157-161

Dragunow M, Faull RLM (1990) MK-801 induces c-fos protein in thalamic and neocortical neurons of rat brain. Neurosci Lett 113:144-150

Dragunow M (1992) Axotomized medial septal-diagonal band neurons express Jun-like immunoreactivity. Mol Brain Res 15:141-144

Dragunow M, Robertson HA (1988) Brain injury induces c-fos protein(s) in nerve and glial-like cells in adult mammalian brain. Brain Res 455:295-299

Dragunow M, Faull RLM, Jansen KLR (1990a) MK-801, an antagonist of NMDA receptors, inhibits injury-induced c-fos protein accumulation in rat brain. Neurosci Lett 109:128-133

Dragunow M, Goulding M, Faull RLM, Ralph R, Mee E, Frith R (1990b) Induction of c-fos mRNA and protein in neurons and glia after traumatic brain injury: pharmacological characterization. Exp Neurol 107:236-248

Dragunow M, de Castro D, Faull RLM (1990c) Induction of fos in glia-like cells after focal brain injury but not during wallerian degeneration. Brain Res 527:41-54

Dragunow M, Yamada N, Bilkey DK, Lawlor P (1992) Induction of immediate-early gene proteins in dentate granule cells and somatostatin interneurons after hippocampal seizures. Mol Brain Res 13:119-126

Dragunow M, Currie RW, Faull RLM, Robertson HA, Jansen K (1989b) Immediate-early genes, kindling and long-term potentiation. Neurosci Biobehav Rev 13 (4):301-313

Dragunow M (1990) Presence and induction of Fos B-like immunoreactivity in neural, but non-neural, cells in adult rat brain. Brain Res 553:324-328

Dragunow M, Faull RLM (1989) Rolipram induces c-fos protein-like immunoreactivity in ependymal and glial-like cells in adult rat brain. Brain Res 501:382-388

Dragunow M, Currie RW, Robertson HA, Faull RLM (1989c) Heat shock induces c-fos protein-like immunoreactivity in glial cells in adult rat brain. Exper Neurol 106:105-109

Dragunow M, Young D, Hughes P, MacGibbon G, Lawlor P, Singleton K, Sirimanne E, Beilharz E, Gluckman P (1993) Is c-jun involved in nerve cell death following status epilepticus and hypoxic-ischaemic brain injury? Brain Res Mol Brain Res 18:347-52

Dragunow M, Hughes P (1993) Differential expression of immediate-early proteins in non-nerve cells after focal brain injury. Int J Devl Neurosci 11:249-55

Dragunow M, Robertson HA, Robertson GS (1988) Amygdala kindling and c-fos protein(s). Exp Neurol 102:261-263

Dragunow M, Robertson GS, Faull RLM, Robertson HA, Jansen K (1990c) D_2 dopamine receptor antagonists induce fos and related proteins in rat striatal neurons. Neurosci 37:287-294

Dragunow M, Logan B, Laverty R. (1991b) 3,4-Methylenedioxymethamphetamine induces Fos-like proteins in rat basal ganglia: reversal with MK 801. Eur J Pharmacol 206:255-258

Dragunow M, Leah JD, Faull RLM (1991a) Prolonged and selective induction of fos-related antigen(s) in striatal neurons after 6-hydroxydopamine lesions of the rat substantia nigra pars compacta. Mol Brain Res 10:355-358

Dragunow M, Peterson MR, Robertson HA (1987) Presence of c-fos-like immunoreactivity in the adult rat brain. Eur J Pharmacol 135:113-114

Ernfors P, Bengzon J, Kokaia Z, Persson H, Lindvall O (1991) Increased levels of messenger RNAs for neurotrophic factors in the brain during kindling epileptogenesis. Neuron 7:165-176

Evan GI, Wyllie AH, Gilbert CS, Littlewood TD, Land H, Brooks M, Waters CM, Penn LZ, Hancock DC (1992) Induction of apoptosis in fibroblasts by c-myc protein. Cell 69 1:119-128

Faden AI, Demediuk P, Scott Panter S, Vink R (1989) The role of excitatory amino acids and NMDA receptors in traumatic brain injury. Science 244:798-799

Falkenberg T, Mohammed AK, Henriksson B, Persson H, Winblad B, Lindefors N (1992) Increased expression of brain-derived neurotrophic factor mRNA in rat hippocampus is associated with improved spatial memory and enriched environment. Neurosci Lett 138:153-156

Fanidi A, Harrington EA, Evan GI (1992) Cooperative interaction between c-*myc* and *bcl*-2 proto-oncogenes. Nature 359:554-556

Funder J (1993) Mineralocorticoids, glucocorticoids, receptors and response elements. Science 259:1132-1133

Gall CM, Isackson PJ (1989) Limbic seizures increase neuronal production of messenger RNA for nerve-growth factor. Science 245:758-761

Garcia I, Martinou I, Tsujimoto Y, Martinou J-C (1992) Prevention of programmed cell death of sympathetic neurons by the *bcl*-2 proto-oncogene. Science 258:302-304

Gluckman P, Klempt N, Guan J, Mallard C, Sirimanne E, Dragunow M, Klempt M, Singh K, Williams C, Nikolics K. A role for IGF-1 in the rescue of CNS neurons following hypoxic-ischaemic injury. Biochem Biophys Res Comm 182 (2):593-599

Goc A, Norman SA, Puchacz E, Stachowiak EK, Lukas RJ, Stachowiak MK (1992) A 5'-Flanking region of the bovine tyrosine hydroxylase gene is involved in cell-specific expression, activation of gene transcription by phorbol ester, and transactivation by c-fos and c-jun. Mol and Cell Neurosci 3:383-394

Goto K, Ishige A, Sekiguchi K, Iizuka S, Sugimoto A, Yuzurihara M, Aburada M, Hosoya E, Kogure K (1990) Effects of cycloheximide on delayed neuronal death in rat hippocampus. Brain Res 534:299-302

Gunn AJ, Dragunow M, Faull RLM, Gluckman PD (1990) Effects of hypoxia-ischemia and seizures on neuronal and glial-like c-*fos* protein levels in the infant rat. Brain Res 531:105-116

Hai T, Curran T (1991) Cross-family dimerization of transcription factors Fos/Jun and ATF/CREB alters DNA binding specificity. Proc Natl Acad Sci USA 88:3720-3724

Hengerer B, Lindholm D, Heumann R, Rüther U, Wagner EF, Thoenen H (1990). Lesion-induced increase in nerve growth factor mRNA is mediated by c-*fos*. Proc Natl Acad Sci USA 87:3899-3903

Hughes P, Young D, Dragunow M (1993a) MK-801 sensitizes rats to pilocarpine induced limbic seizures and status epilepticus. Neuroreport 4: 314-316

Hughes P, Dragunow M (1993) Muscarinic receptor mediated induction of Fos protein in rat brain. Neurosci Lett 150:122-126

Hughes P, Lawlor P, Dragunow M (1992) Basal expression of Fos, Fos-related, Jun, and Krox-24 proteins in rat hippocampus. Mol Brain Res 13:355-357

Hughes P, Dragunow M, Beilharz E, Lawlor P, Gluckman P (1993b) MK-801 induces immediate-early gene proteins and BDNF mRNA in rat cerebrocortical neurons. Neuroreport 4:183-186

Hughes P, Beilharz E, Gluckman P, Dragunow M (1993c) Brain-derived neurotrophic factor is induced as an immediate-early gene following N-methyl-D-aspartate receptor activation. Neuroscience 57:319-328

Hunt SP, Pini A, Evan G (1987) Induction of c-fos-like protein in spinal cord neurons following sensory stimulation. Nature 328:632-634

Hunter T, Karin M (1992) The regulation of transcription by phosphorylation. Cell 70:375-387

Jeffery KJ, Abraham WC, Dragunow M, Mason SE (1990) Induction of fos-like immunoreactivity and the maintenance of long-term potentiation in the dentate gyrus of unanesthetized rats. Mol Brain Res 8:267-274

Joseph R, Li W, Han E (1993) Neuronal death, cytoplasmic calcium and internucleosomal DNA fragmentation: evidence for DNA fragments being released from cells. Mol Brain Res 17:70-76

Kerpolla TK, Curran T (1991) Fos-Jun heterodimers and jun homodimers bend DNA in opposite directions: implications for transcription factor cooperativity. Cell 66:317-326

Koh J-Y, Cotman CW (1992) Programmed cell death: its possible contribution to neurotoxicity mediated by calcium channel antagonists. Brain Res 587:233-240

Kovary K, Bravo R (1991) Expression of different jun and fos proteins during the G_0-to-G_1 transition in mouse fibroblasts: in vitro and in vivo associations. Mol & Cell Biol 11:2451-2459

Kromer LF (1986) Nerve growth factor treatment after brain injury prevents neuronal death. Science 235:214-216

Leah JD, Herdegen T, Dragunow M, Murahov A, Bravo R (1993) Expression of immediate-early gene proteins following axotomy and inhibition of axonal transport in the rat CNS. Neuroscience 57:53-66

Lerea LS, Butler LS, McNamara JO (1992) NMDA and Non-NMDA Receptor-mediated increase of c-fos mRNA in dentate gyrus neurons involves calcium influx via different routes. J Neurosci 12(8):2973-2981

MacGibbon G, Lawlor P, Bravo R, Dragunow M (1994) Clozapine and haloperidol produce a differential pattern of immediate-early gene expression in rat caudate-putamen, nucleus accumbens, lateral septum and islands of calleja. Mol Brain Res 23:21-32

Martin DP, Schmidt RE, DeStefano PS, Lowry OH, Carter JG, Johnson EM Jr. (1988) Inhibitors of protein synthesis and RNA synthesis prevent neuronal death caused by nerve growth factor deprivation. J Cell Biol 106:829-844

Moochetti I, De Bernardi MA, Szekely AM, Alho H, Brooker G, Costa E (1989) Regulation of nerve growth factor biosynthesis by beta-adrenergic receptor activation in astrocytoma cells: a potential role of c-*fos* protein. Proc Natl Acad Sci USA 86:3891-3895

Morgan JI, Cohen DR, Hampstead JL, Curran T (1987) Mapping patterns of c-fos expression in the CNS after seizure. Science 237:192-197

Morgan JI, Curran T (1991) Stimulus-transcription coupling in the nervous system: involvement of the inducible proto-oncogenes *fos* and *jun*. Ann Rev Neurosci 14:421-451

Nakabeppu Y, Nathans D (1991) A naturally occurring truncated form of FosB that inhibits Fos/Jun transcriptional activity. Cell 64: 751-759

Najilerahim A, Showell DGL, Pearson RCA (1991) Transient increase in glutamic acid decarboxylase mRNA in the cerebral cortex following focal cortical lesion in the rat. Exp Brain

Res 87:113-118

Olenik C, Lais A, Meyer DK (1991) Effects of unilateral cortex lesions on gene expression of rat cortical cholecystokinin neurons. Mol Brain Res 10:259-265

Papas S, Crépel V, Hasboun D, Jorquera I, Chinestra P, Ben-Ari Y (1992) Cycloheximide reduces the effects of anoxic insult *in vivo* and *in vitro*. Eur J Neurosci 4:758-765

Pulverer BJ, Kyriakis JM, Avruch J, Nikolakaki E, Woodgett JR (1991) Phosphorylation of c-jun mediated by MAP kinases. Nature 353:345-349

Riabowol KT, Vosatka RJ, Ziff EB, Lamb NJ, Feramisco JR (1988) Microinjection of fos-specific antibodies blocks DNA synthesis in fibroblast cells. Mol Cell Biol 8:1670-1676

Richardson CL, Tate WP, Mason SE, Lawlor PA, Dragunow M, Abraham WC (1992) Correlation between the induction of an immediate early gene, *zif*/268, and long-term potentiation in the dentate gyrus. Brain Res 580:147-154

Robertson GS, Herrera DG, Dragunow M, Robertson HA (1989) L dopa activates c-fos in the striatum ipsilateral to a 6-hydroxydopamine lesion of the substantia nigra. Eur J Pharmacol 159:99-100

Rosen JB, Cain CJ, Weiss S, Post RM (1992) Alterations in mRNA of enkephalin, dynorphin and thyrotropin releasing hormone during amygdala kindling: an in situ hybridization study. Mol Brain Res 15:247-255

Ruppert C, Wille W (1987) Proto-oncogene c-fos is highly induced by disruption of neonatal but not of mature brain tissue. Mol Brain Res 2:51-56

Saffen DW, Cole AJ, Worley PF, Christy BA, Ryder K, Baraban JM (1988) Convulsant-induced increase in transcription factor messenger RNAs in rat brain. Proc Natl Acad Sci USA 85:7795-7799

Schreiber SS, Najm I, Tocco G, Baudry M (1992) Cycloheximide treatment prevents neurotoxicity and c-*fos* expression in adult rat brain following kainic acid treatment. Society for Neurosc, Abstracts 18:81

Schüle R, Rangarajan P, Yang N, Kliewer S, Ransone LJ, Bolado J, Verma IM, Evans RM (1991) Retinoic acid is a negative regulator of AP-1-responsive genes. Proc Natl Acad Sci USA 88:6092-6096

Seyfert VL, McMahon SB, Glenn WD, Yellen AJ, Sukhatme VP, Cao X, Monroe JG (1990) Methylation of an immediate-early inducible gene as a mechanism for B Cell tolerance induction. Science 250:797-799

Shi Y, Glynn JM, Guilbert LJ, Cotter TG, Bissonnette RP, Green DR (1992) Role for c-myc in activation-induced apoptotic cell death in T cell hybridomas. Science 257:212-217

Shigeno T, Yamasaki Y, Kato G, Kusaka K, Mima T, Takakura K, Graham DI, Furukawa S (1990) Reduction of delayed neuronal death by inhibition of protein synthesis. Neurosci Lett 120:117-119

Shigeno T, Mima T, Takakura K, Graham DI, Kato G, Hashimoto Y, Furukawa S (1991) Amelioration of delayed neuronal death in the hippocampus by nerve growth factor. J Neurosci 11(9):2914-2919

Sonnenberg JL, Rauscher FJ (III), Morgan JI, Curran T (1989) Regulation of proenkephalin by Fos and Jun. Science 246:1622-1624

Sutula T, Cavazos J, Golarai G (1992) Alteration of long-lasting structural and functional effects of kainic acid in the hippocampus by brief treatment with phenobarbital. J Neurosci 12(11):4173-4187

Weiser M, Baker H, Wessel TC, Joh TH (1993) Axotomy-induced differential gene induction in neurons of the locus ceruleus and substantia nigra. Mol Brain Res 17:319-327

Wetmore C, Cao Y, Petterson RF, Olson L (1991) Brain-derived neurotrophic factor: Subcellular compartmentalization and interneuronal transfer as visualized with anti-peptide antibodies. Proc Natl Acad Sci USA 88:9843-9847

Williams J, Dragunow M, Lawlor P, Mason B, Abraham W, Leah J, Bravo R, Demmer J, Tate W (1994) Krox 20 may play a key role in the stabilization of long-term potentiation. Mol Brain Res (in press)

White JD, Gall CM (1987) Differential regulation of neuropeptide and proto-oncogene mRNA content in the hippocampus following recurrent seizures. Mol Brain Res 3:21-29

Xanthoudakis S, Curran T (1992) Identification and characterization of Ref-1, a nuclear protein that facilitates AP-1 DNA-binding activity. EMBO J 11:653-665

Young D, Dragunow M (1993) Non-NMDA glutamate receptors are involved in the maintenance of status epilepticus. Neuroreport 5:81-83

Zhang X-K, Wills KN, Husmann M, Hermann T, Pfahl M (1991) Novel pathway for thyroid hormone receptor action through interaction with jun and fos oncogene activities. Mol and Cell Biol 11:6016-6025

Immediate-early genes in nociception

T.R. Tölle [a], J. Schadrack [a], J.M. Castro-Lopes [b],
and W. Zieglgänsberger [a].
[a] Max-Planck-Institute of Psychiatry, Clinical Institute, Clinical
Neuropharmacology, D-80804 München (Germany) and [b] Institute of Histology
and Embryology, Faculty of Medicine of Oporto, 4200 Porto (Portugal)

Introduction

Cellular immediate-early genes (IEGs) share a close structural homology with some viral oncogenes. In the eukaryotic genome these counterparts of oncogenes have been termed proto-oncogenes. Recent advances in cellular biology have identified the activation and deactivation of IEGs as molecular mechanisms to control regulated and deregulated growth, cellular differentiation and development. In neurobiology IEGs are believed to be involved in the neuron's ability to convert short-term synaptic stimulation into long-lasting responses and thus contributing to the adaptive alterations involved in neuronal plasticity (Evan 1991; Goelet et al. 1986; for review see Morgan and Curran 1991). IEGs may be viewed as "third messengers" in a stimulus-transcription cascade transferring extracellular information via second messenger systems, such as calcium, into changes in target-gene transcription, thereby changing the phenotype of neurons. IEGs which encode proteins with a leucine zipper structure have to dimerize to function. While Jun proteins can form homo- as well as heterodimers, members of the Fos family require dimerization with Jun proteins. The various IEG complexes possess different binding affinities for DNA consensus sequences (AP-1 site) and variable transcriptional activities. The zinc-finger protein Krox-24, also termed NGFI-A, Egr-1, Zif/268, can achieve transcriptional activation in a monomeric fashion.

The expression of IEGs has been widely exploited to indicate the activation of neurons with single cell resolution and to perform neuroanatomical tract tracing approaches. Recent advances in pain research illustrate the analytical power of modern neurosciences in a field previously accessible only to methods of systems biology. Novel molecular and cellular biological techniques have changed the face of pain research by detailing the multiplicity of pain transducing and pain suppressive systems. The renaissance of concepts of neuronal plasticity in the field of pain was substantially triggered by the observation of Hunt et al. (1987) that sensory stimulation can transsynaptically induce the expression of IEGs and has led to headlines such as "c-Fos and the changing face of pain" (Fitzgerald 1990). The experimental data from basic research are beginning to shed new light on the explanation of clinical observations in man such as hyperalgesia, allodynia, phantom limb pain and the pathogenesis of chronic pain. Novel compounds and new regimens for drug treatment to prevent activity-dependent long-term changes or to facilitate extinction in pain-related systems are now beginning to emerge.

The perception of painful stimuli involves various levels of the neuraxis from the periphery to the cortex; each of these levels probably receives and is the origin of modulatory neuronal mechanisms. Multiple neurotransmitter and -modulator systems participate in the integration of nociceptive afferents (Aδ- and C-fibers) in the dorsal horn of the spinal cord. Excitatory and inhibitory amino acid transmitters, monoamines and various neuropeptides have been demonstrated in primary afferents, intrinsic spinal neurons, ascending as well as descending pathways (for review see Zieglgänsberger 1986; Zieglgänsberger and Tölle 1993). Here we focus on the physiology and pharmacology of IEG expression in the spinal cord following somatic and visceral noxious stimulation. Depending on the individual IEG studied (c-Fos, Fos B, c-Jun, Jun B, Jun D or Krox-24/NGFI-A) and the noxious stimulus applied (mechanical, thermal, chemical) the pattern of expression shows considerable differences in terms of the number of neurons, the intraspinal anatomical distribution, the peak time of expression and the steady-state persistance of the IEG signal. The pharmacological studies concentrate on opioidergic mechanisms by testing the modulatory action of morphine, kelatorphan (an inhibitor of multiple enkephalin-degrading enzymes) and naloxone. Additionally, the effects of drugs modulating glutamatergic neurotransmission (Ketamine and MK-801 as non-competitve NMDA-antagonists) and the effects of anticonvulsant drugs are presented.

Methods

Experiments were performed in continously anesthetized rats or cats. Chronic monoarthritis was induced according to Butler et al. (1992). The different kinds of noxious stimulation are briefly presented in the corresponding paragraphs. For detection of the different IEG encoded proteins, immunocytochemical procedures were performed with polyclonal antibodies that were generated in rabbits immunized with bacterially expressed fusion proteins of the particular IEG protein. Animals were perfused and fixated at the various time points depending on the kind of noxious stimulus applied. After the respective neural tissues had been processed for immunocytochemistry, the immunostaining reaction was performed using an avidin-biotin complex method (Vectastain, Vector Laboratories, U.S.A.). For data analysis and more detailed information on the experimental procedures we refer to the respective publications.

IEG Expression following noxious and non-noxious stimuli

Non-noxious stimulation

The expression of c-Fos and other IEGs in the central nervous system has been described for different types of noxious and non-noxious stimulation. In the rat spinal cord brushing of the hairs of the hind foot or non-noxious manipulation of the hind limb induced expression of c-Fos whereas following stimulation with radiant heat of 40°C (a temperature considered as non-noxious) no c-Fos immunoreactivity (IR) was detected in the spinal cord, in the dorsal root ganglion cells or in the nucleus gracilis (Hunt et al. 1987). Rats in conditions of forced motoric stimulation (forced running in a wheel for 1 hour) showed c-Fos IR in the cervical and lumbar enlargements of the spinal cord, as well as in the nuclei of the dorsal column, in the nucleus cuneatus lateralis and, in a high number, in the cerebellum, whereas noxious stimulation did not induce c-Fos in these three latter areas (Gogas et al. 1990). Dorsal rhizotomie of the roots C5-Th1 markedly reduced the IR following forced running, thus suggesting that synaptic activation through fibers with a large diameter, deriving from muscles or joints, was the origin of c-Fos induction (Gogas et al. 1990). Expression of c-Fos was also observed in the spinal cord following ligature of the sciatic nerve leading to the development of neuroma (Chi et al. 1993a, 1993b). Nerve cut of the trigeminal nerve resulted in some c-Fos IR in laminae I and II of the trigeminal nucleus (Sharp et al. 1989).

Fos and Jun proteins, as well as Krox-24, were differentially induced in rat dorsal horn neurons of segments L4-5 by transcutaneous electrical stimulation of the sciatic nerve. Electrical stimulation of Aδ- and C-fibers, but not of Aα- and Aβ-fibers, induced all IEGs (Herdegen et al. 1991a).

Acute noxious stimulation

In the spinal cord of unstimulated rats, expression of c-Fos, Fos B, c-Jun and Jun B was only detected in a few neurons and c-Jun also showed staining in some motoneurons. Krox-24/NGFI-A and Jun D showed a basal expression in 10-20 neurons in the superficial and deep laminae of the dorsal horn (25μm transversal section). The basal expression of the various IEGs was not influenced by the various anesthetics. This is in accordance with other reports concerning IEG expression in the spinal cord. However, some authors (Bullitt 1989; Presley et al. 1990) mention a diminution due to anasthesia.

In general, following acute somatic noxious stimulation, IEGs are expressed in the superficial layers of the dorsal horn (lamina I and II; sDH) and the deeper laminae (III-VII and X; dDH), but rarely in the ventral horn of the spinal cord (Herdegen et al. 1991a; Hunt et al. 1987, see this volume; Menetrey et al 1989; Tölle et al. 1990, 1994a).

Following noxious heat stimulation (immersion of one hind paw in 52°C hot water 10 x for 20 sec, with a 90 sec interval) almost all the immunoreactive neurons for the IEG-encoded proteins of each kind were found in the ipsilateral dorsal horn with a more pronounced expression in the medial half of the sDH. The highest amount of immunoreactive neurons was found in the lumbar segments $L_{4/5}$ (about 100 neurons/ section) and extended rostrocaudally from the spinal segments L_2-S_1, in agreement with the known somatotopical organization. IEG expression extended rostrocaudally over 3-4 segments regarding the superficial layers, whereas, in the deeper laminae, a higher number of segments was affected (Herdegen et al. 1991a; Hunt et al. 1987; Tölle et al. 1990). The spatial pattern of expression (rostrocaudal and laminar) is in agreement with anatomical tracing studies describing the projections of primary afferent fibers to the spinal cord (Molander et al. 1984; Sweet and Woolf 1985). This distribution of IEG expressing neurons is also in accordance with electrophysiological data showing that regions with IEG positive neurons correspond to regions were neurons are activated by nociceptive stimulation (for refs see Besson and Chaouch 1987; Willis and Coggeshall 1991; Zieglgänsberger 1986). Bullitt et al. (1991) used c-Fos mapping as "activity markers" to study the somatotopical distribution of spinal cord neurons following noxious stimulation of individual digits of the foot; findings which also had been reported by Williams et al. (1989). On the level of the trigeminal system Strassman and Vos (1993) detailed, with c-Fos mapping, the differential activation of trigeminal neurons following various types of facial stimulations.

In the spinal cord different types of somatic stimuli evoke a differential pattern of c-Fos expression (Lima et al. 1993). Chemical stimulation by subcutaneous injection of 20% formalin resulted in a preferential activation of c-Fos in lamina I neurons, which accounted for approximately 60% of the total number of neurons showing Fos IR. Thermal stimulation by radiant heat (65°C) caused a marked expression of c-Fos in laminae I and IIo (ca. 95%), but mechanical stimulation, by pinching or needle prick, elicited a wider distribution of c-Fos positive neurons, which was about 25% in lamina I and 10-20% in the laminae II-V.

Visceral nociception was studied with several types of stimulation, such as i.p. injection of acetic acid, which induced c-Fos IR at the thoraco-lumbar junction (Th_{12}-L_2) (DeLeo et al. 1991; Hammond et al. 1992; Menetrey et al. 1989), colorectal distension (Traub et al. 1993), electric stimulation of the pelvic nerve (Birder et al. 1991) or vaginocervical stimulation (Chinapen et al. 1992). Visceral stimulation has also been provoked by chemical and mechanical irritation of the urinary tract in the rat (Birder and deGroat 1992a). c-Fos expression following the latter types of stimulation expanded rostrocaudally over spinal cord segments L_6-S_3 and was localized in the superficial and deeper laminae, in correspondence to the somatotopical organisation of the primary afferent fibers originating in the viscera. Vesical stimulation in the cat (unpublished observations together with C. Vahle-Hinz) (Fig. 1) induced c-Fos expression in the lateral parts of the superficial dorsal horn and in the dorsal commissure. In the rat the hypogastric and pelvic nerve are known to carry primary afferents from the bladder. Interestingly, in the hypogastric area most of the neurons were located in lamina I, which is in contrast to the spinal distribution of Fos positive neurons in the pelvic area, where it included, besides lamina I, the dorsal grey commissure and the intermediolateral grey matter (Cruz et al. 1993). Neonatal capsaicin dramatically reduced the expression of c-Fos in all regions of both areas (Cruz et al. submitted), whereas low doses of intravesical capsaicin reduced predominantly c-Fos expression in the area of pelvic nerve afferents, particularly in lamina I.

Temporo-spatial pattern

The expression of c-Fos, Fos B, c-Jun, Jun B, Jun D and Krox24/NGFI-A, following peripheral noxious thermal stimulation, showed that several transcription

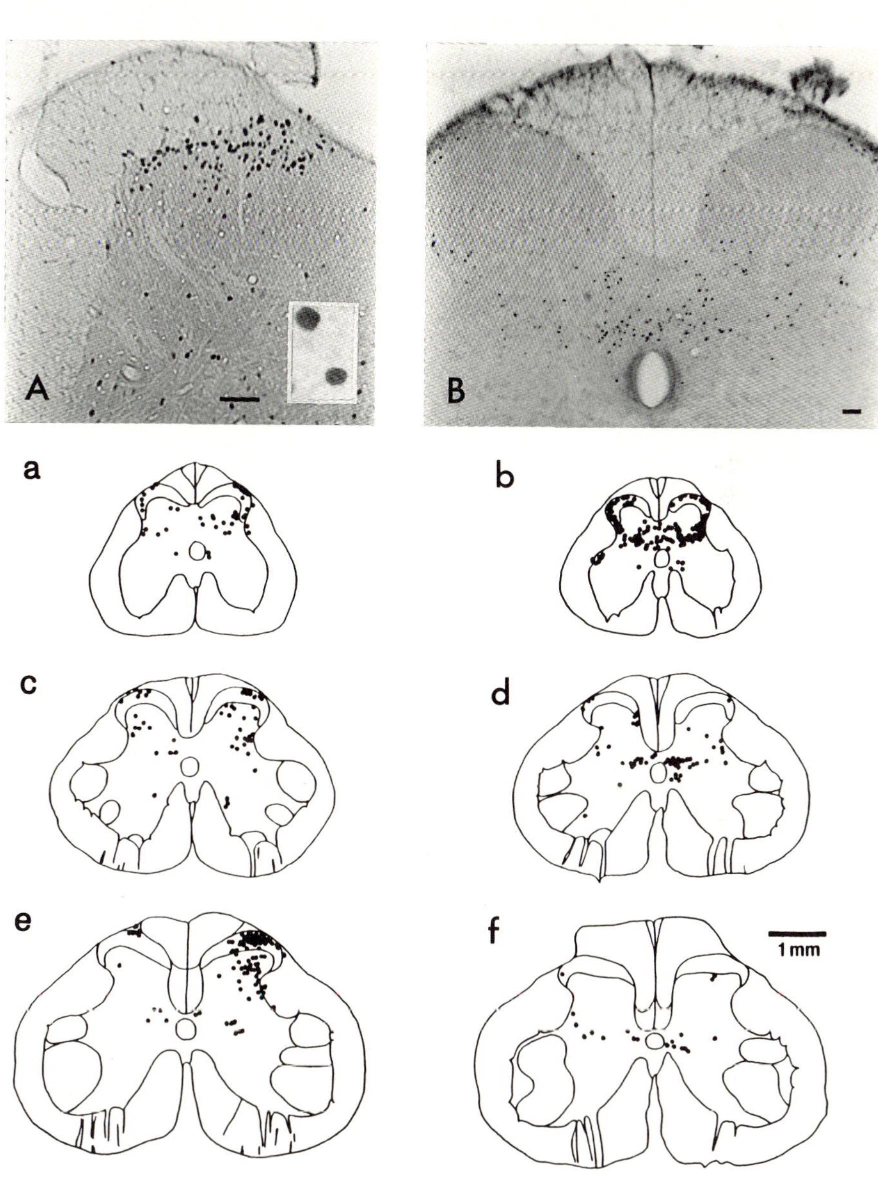

Fig. 1. Photomicrogaphs and schematic illustration of the distribution of c-Fos positive neurons in the rat (A,a,c,e) or the cat (B,b,d,f) spinal cord following noxious somatic (thermal, 10 x 52°C, 20s on, 90s off) stimulation of one hind paw or visceral stimulation of the bladder (pressure of 100 mm Hg). The photomicrographs and the schematic levels (dot per neuron) show all labelled neurons found in a representative 25μm (rat) or 40μm (cat) slice of the appropriate spinal level. Note the highest labelling in the $L_{4/5}$ region or $S_{1/2}$ region, respectively. (Scale bar in A,B = 100 μm). The inset in higher magnification exemplifies the typical immunostaining of the cell nucleus sparing the nucleolus.

factors are induced with a distinct temporo-spatial pattern in the spinal cord (Herdegen et al. 1991a, 1991b; Tölle et al. 1994a). The first increase of the Fos protein in sDH was detected 30 min after acute noxious stimulation, and maximal expression was seen after 2 hours (Fig. 1). In the deep dorsal horn, the expression of c-Fos is delayed and peak expression is detected 2-4 hours after stimulation. Thereafter, the number of neurons expressing c-Fos decrease and, after 24 hours, are not detectable (Herdegen et al. 1991b; Tölle et al. 1994a; Tölle et al. unpublished observations). The maximal staining in the sDH after 2 hours for c-Fos, c-Jun, Jun B and Krox-24/NGFI-A confirms the initial finding of Hunt et al. (1987) concerning c-Fos and is in concordance with an increase in NGFI-A, NGFI-B, c-*fos*, SRF and c-*jun* mRNAs following a single noxious heat stimulus (Wisden et al. 1990), as well as a specific spatial and temporal pattern of Jun B, Jun D and Fos B IR following repetitive noxious heat stimuli or C-fiber stimulation (Herdegen et al. 1991a, 1991b). In the dDH the time course of expression was comparable to the sDH, but at a lower level. c-Fos, Krox-24, c-Jun and Jun B peaked after 2 hours and steadily declined for the following 6 hours to relative values ranging between 10-50% of maximum expression. For Fos B the number of immunoreactive neurons continously increased in sDH and dDH for the entire observation period of 8 hours. Jun D demonstrated a delayed onset of expression in both sDH and dDH and showed a first increase only 4 hours after termination of noxious heat stimulation which was still present 4 hours later (Fig. 2). Whereas c-Fos, Jun D and Krox-24 were expressed in numerous nuclei in the dDH, c-Jun, Jun B and Fos B were almost absent.

Acute inflammation induced by subcutaneous injection of formalin into one hind paw caused expression of c-Fos and Krox-24/NGFI-A in sDH and dDH of the rat spinal cord lumbar segments (Fig. 6) (see also Menetrey et al. 1989; Pertovaara et al. 1993; Presley et al. 1990), and in cervical segments following injection into the forepaw (Abbadie et al. 1992b). Compared to noxious heat stimulation, acute inflammation showed more intense staining in neurons in the deep laminae of the spinal cord, such as laminae IV and V. Injection of formalin induced a rapid onset and rapid decrease of Jun B expression, whereas Fos B and Jun D showed a delayed expression and a longer persistence (Herdegen et al. 1991b). On the contralateral side of stimulation a non-significant 5-10% enhancement of IEG expression was detected following noxious heat stimulation (Tölle et al. 1994a). A 'second wave' of weak bilateral c-Fos expression was reported in the deep dorsal horns after noxious skin heating (Williams et al. 1990; see also Hunt et al. this volume). In addition, sensitization of dorsal horn neurons was reported in studies of successive acute mechanical or chemical stimulation, showing that the expression of Fos proteins was augmented by an additional prior contralateral stimulus in comparison to a single stimulus (Leah et al. 1992).

With regard to the IEG c-*fos*, one hypothesis is that the amount of IEG protein produced by dorsal horn neurons is proportional to the degree of synaptic excitation and, that the expression of c-Fos depends upon the neurochemistry of synaptic transmission, and/or upon the postsynaptic propensity for its induction (Bullitt et al. 1992; Hunt et al. 1987; Williams et al. 1989). The diversity of chemical neuroanatomy for neurotransmitters and -modulators in the dorsal horn (for refs

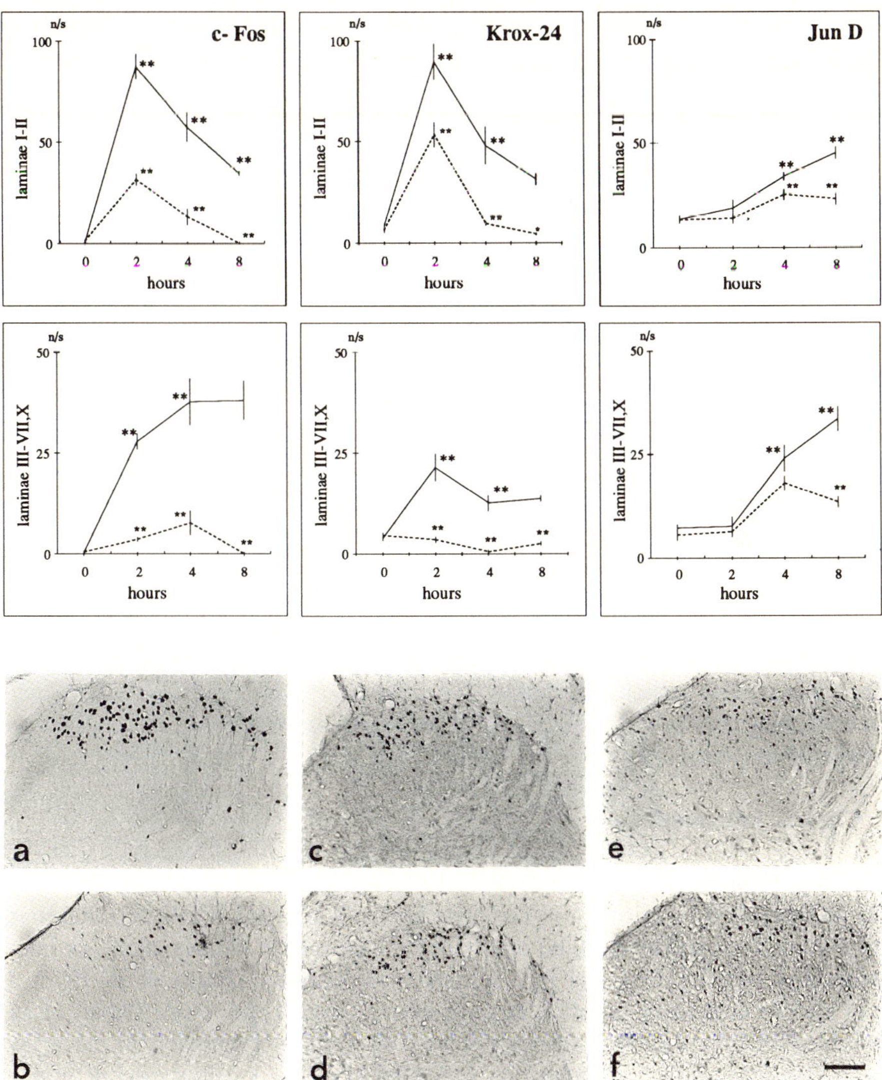

Fig. 2. Time course of expression of c-Fos, Krox-24 and Jun D proteins in laminae I-II (sDH) and laminae III-VII,X (dDH) neurons in the ipsilateral rat spinal cord following noxious thermal stimulation of one hind paw (basal expression=0hours, 2h, 4h, 8h; closed line) and modulation by prior application of intravenous morphine (10 mg/kg; dashed line). Values given represent mean±S.E.M. of labeled neurons in a single 25 µm slice in the lumbar segments L_4-L_5 expressed as neurons/section (n/s). Note the halved scaling of the y-axis for neurons of deep laminae. One or two asterisks indicate statistical significance (p<0.05; p<0.01) in Duncan *post hoc* comparisons that were separately performed for sDh and dDH between control animals following noxious stimulation (2 vs 0h, 4 vs 2h, 8 vs 4h; symbols: *,**) and control vs morphine animals (0,2,4,8h; symbols *,**).

The additional photomicrographs illustrate the distribution of c-Fos, Krox-24 and Jun D immunostained neurons in the lumbar spinal cord. The samples depicted show the time points of maximum expression of c-Fos, Krox-24 or Jun D (a,c,e) following noxious thermal stimulation of one hind paw and prior application of morphine (10 mg/kg; b,d,f). (Scale bar = 100 µm).

stimulation in the cat (Matsumoto et al. 1993) a dose-related effect of opioids on IEG induction could be demonstrated.

Besides the substantial reduction of IEG expression in the spinal cord (Hammond et al. 1992; Presley et al. 1990; Tölle et al. 1990), opiates also caused a dose-dependent inhibition of pain behaviour (Presley et al. 1990) and a suppres-

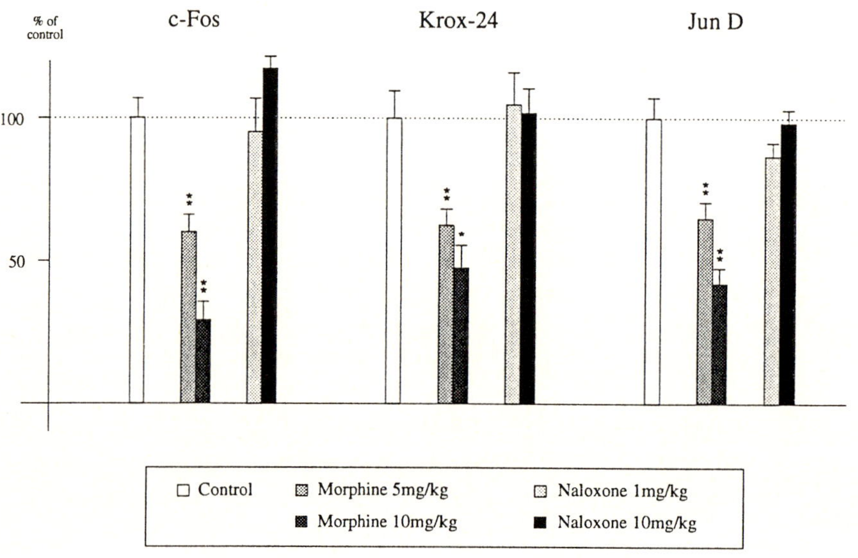

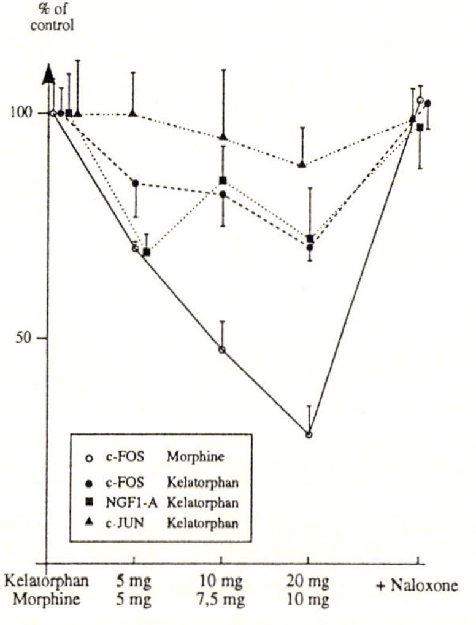

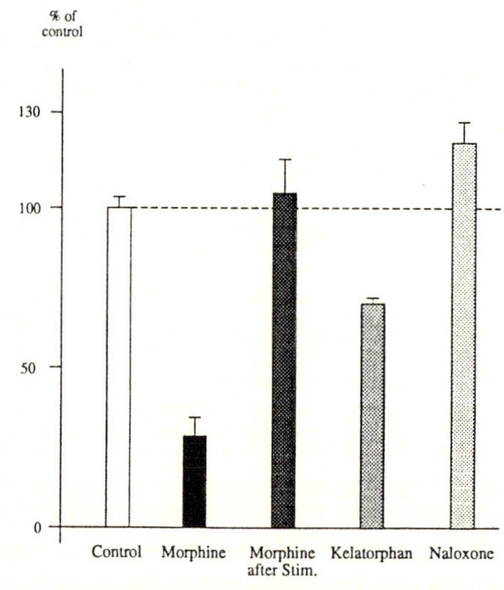

sion of noxious stimulus-evoked Fos IR in more rostral structures (Hammond et al. 1992). The authors concluded (Gogas et al. 1991; Hammond et al. 1992; Presley et al. 1990) that production of complete anti-nociception is not associated with a total suppression of c-Fos. Alternatively, behavioural analgesia might be achieved without complete anti-nociception, at least at the spinal cord level. The depression on neuronal discharge activity of opioid agonists is well documented (Gogas et al. 1991; Hammond et al. 1992; Presley et al. 1990; Tölle et al. 1990), acting directly at the spinal cord level (Yaksh 1981) or indirectly, by activation of supraspinal monaminergic inhibitory control systems (Basbaum and Fields 1984; LaMotte et al. 1976; for review see Zieglgänsberger 1986). Similar to pretreatment with morphine, electrical stimulation in the nucleus raphé magnus reduced c-Fos IR in dorsal horn neurons in response to peripheral noxious heat stimulation (Jones and Light 1990). However, the contribution of descending pathways to the pattern of IEG expression following noxious stimulation (Bullitt 1991) seems to be minor, as revealed in studies with spinally transsected animals (Gogas et al. 1991). Extra- and intracellular studies show that morphine preferentially supresses C-fiber evoked activity, whereas early components of the response resulting from the activation of large myelinated fibers, are affected only slightly if at all even by high doses of morphine (Besson and Chaouch 1987; Dickenson and Sullivan 1986; Zieglgänsberger 1986). Thus the residual expression of IEGs, still observed under morphine, might be explained in the neuronal activity evoked by sustained synaptic bombardment, and/or excitation derived from low-threshold afferents activated during noxious stimulation. Whether a more pronounced liability to the depressant effect of morphine or a distinct propensity of neurons located in deeper laminae (Presley et al. 1990; Tölle et al. 1990) to express IEGs is the reason why they are more readily affected by opioids remains to be shown. Distinct intracellular pathways that initiate or control the start and the persistence of Fos B and Jun D expression could account for their comparatively low sensitivity to morphine.

← **Fig. 4.** The upper histogramm shows the expression of IEG (c-Fos, Krox-24 and Jun D) positive neurons in laminae I to VII and X of the rat spinal cord following noxious thermal stimulation and pre-treatment with increasing doses of morphine (5 and 10 mg/kg). As a control for morphine effects animals were injected with naloxone (1 or 10 mg/kg) before application of morphine (10 mg/kg). The effects of morphine and naloxone were investigated at time points where the various IEGs showed maximal expression following termination of noxious stimulation (2h for c-Fos, Krox-24; 8h for Jun D). Values given represent mean+S.E.M of neurons per section relative to the corresponding control values. Asterisks indicate statistical significances (* $p<0.05$; ** $p<0.01$) in Duncan *post hoc* tests following a one-way ANOVA. Significance for 5 mg/kg morphine is expressed as compared to control, and for 10 mg/kg morphine is expressed as compared to 5 mg/kg morphine.
The lower left histogram shows the expression of c-Fos, NGFI-A and c-Jun positive neurons in the dorsal horn of the rat spinal cord following noxious thermal stimulation and pre-treatment with morphine (5,7.5,10 mg/kg), Kelatorphan (5,10,20 mg/kg) and prior application of naloxone (10 mg/kg). The lower right histogram shows the expression of c-Fos positive neurons in the rat spinal cord dorsal horn following noxious thermal stimulation and application of morphine (10 mg/kg) *before* and *after* noxious stimulation, pre-treatment with Kelatorphan (20 mg/kg) and naloxone (10 mg/kg). Values given in the lower histograms represent mean+S.E.M. relative to the corresponding control values (saline injection).

Different metabolic pathways activated by the same combination of neurotransmitters released from small calibre primary afferents following noxious stimulation, may provide variable links to the various members of the AP-1 binding transcription factors (Bartel et al. 1989; Morgan and Curran 1991). This notion is supported by findings concerning IEG expression and long-term potentiation in the hippocampus, as well as IEG expression in the spinal cord following synaptic stimulation, which reveal a selective induction of IEGs as a result of distinct receptor activation (Cole et al. 1989; Kehl et al. 1991; Tölle et al. 1991a).

Naloxone - Kelatorphan

The decrease of IEG induction by opioid agonists is naloxone-reversible (Abbadie and Besson 1993; Abbadie et al. 1994b; Hammond et al. 1992; Presley et al. 1990; Tölle et al. 1990, 1991a, 1994a, 1994b). Administration of naloxone alone reproducibly induced an increase in the number of stained neurons for c-Fos in contrast to the other IEGs tested (Fig. 4). However, although this difference was in the range of 120% of control levels, it was not statistically significant for values of the whole dorsal horn or for the sDH or dDH alone. Others (Abbadie and Besson 1993; Presley et al. 1990) made the same observation of a tendency of increased c-Fos IR after application of 1-2 mg/kg naloxone. A significant increase of c-Fos-like IR due to naloxone could be demonstrated following noxious cold stimulation (Abbadie et al 1994b).

The supposed opioidergic link in the spinal cord (Zieglgänsberger 1980, 1986) is driven synaptically by primary afferents and descending, as well as, propriospinal pathways. The findings support the notion that the efficacy of a segmental and tonically active, inhibitory and opioidergic circuitry, which is presumed to be involved in the processing of nociceptive signals in the spinal cord, is decreased by the opioid antagonist (Gibson et al. 1981; for review see Zieglgänsberger 1986). Inhibitors of enkephalin-degrading enzymes decrease neuronal discharge activity and produce antinociception. (for review see Roques et al. 1993). Application of peptidase inhibitors such as Kelatorphan, an inhibitor of neutral endopeptidase 24.11 (NEP) ("enkephalinase"), aminopeptidase N and dipeptidyl-peptidase, reduces degradation of opioid peptides and, thereby, enhances the tonically active opioidergic circuitry and induces antinociceptive effects (Fournié-Zaluski et al. 1984; Morton et al. 1987; for review see Roques et al. 1993).

In order to further investigate the role of endogenous opioids in the induction of IEGs following noxious stimulation, the expression was studied by inhibiting the catabolism of opioid peptides with Kelatorphan (Tölle et al. 1994b) or RB-101 (Abbadie et al. 1994a).

In noxious heat stimulated animals Kelatorphan had a differential effect for sDH and dDH, reducing the number of c-Fos, Krox-24-/NGFI-A and c-Jun positive neurons in the sDH whereas, in the dDH, a diminution was found only for c-Fos. Kelatorphan reduced IEG expression by 20-30% compared to control animals. Kelatorphan (5, 10, 20 mg/kg) showed no obvious dose-response relationship in the level of suppression (Fig. 4), although a tendency could be found for all IEGs. It is feasible to assume that the dose of Kelatorphan administered was maximal in preventing degradation of the opioid peptides released tonically and/or in response to noxious heat stimulation. As for morphine, combined administration of naloxone and Kelatorphan completely reversed the suppressive effect of the peptidase-inhibitor (Fig. 4).

The high density of enkephalin immunoreactivity (Pretel and Piekut 1991) and the great number of opioid receptors, as well as the particularly high concentrations of NEP ("enkephalinase") in the superficial laminae of the spinal cord (Waksman ct al. 1986), may wcll cxplain why Kclatorphan is particularly effective in suppressing IEG induction in the sDH. Concerning the dDH neurons, the effect of Kelatorphan is most likely indirect and results from an activation of sDH ncurons. Electrophysiological experiments have revealed that Kelatorphan was only able to alter the responses of laminae IV and V neurons when injected into the substantia gelatinosa, but not when directly applied into the vicinity of the soma of these neurons (Morton et al. 1987). Co-administration of Kelatorphan has been shown to markedly potentiate the inhibitory effect of (met5)-enkephalin on discharge activity in the substantia gelatinosa (Morton et al. 1987). Similar to the observations regarding the suppression of IEG induction, a ceiling effect in Kelatorphan-induced analgesic responses was detected.

Anticonvulsant drugs

Anticonvulsants, such as carbamazepine, valproate and phenytoine, are mainly used for treatment of epilepsy but are also therapeutically useful in certain forms of chronic pain, such as trigeminal neuralgia or phantom limb pain, which readily respond to treatment with anticonvulsants (Mumenthaler 1976; Sato and Foong 1983). Regarding modulatory effects on IEG induction, anticonvulsants completely block the expression of c-Fos in the cortex, hippocampus and limbic structures following drug-induced seizures (Dragunow and Robertson 1988; Morgan et al. 1987).

Since the pharmacological profile of anticonvulsants is supposed to involve rather immediate effects as well as long-term effects, such as enzymatic modulation of the GABA system (Macdonald and McLean 1986), the effects following single and repetitive application were investigated (Tölle et al. in press).

Application of carbamazepine, valproate or phenytoine had no effect on basal c-Fos IR. Following noxious heat stimulation carbamazepine differentially reduced the amount of c-Fos positive neurons to 65-78% in the sDH and to 17-32% in the dDH, independently from the frequency of application. Valproate reduced c-Fos expression to 53-76% in sDH and 22-60% in dDH. Additionally for this drug, repetitive application for 4 and 8 days further suppressed c-Fos in the deep dorsal horn neurons. Single application of phenytoine decreased c-Fos expression to 76% of control levels in sDH and to 58% in dDH. In addition to the reduction of the number of neurons, we received the clear impression that carbamazepine, valproic acid and phenytoine-treated animals showed a lighter nuclear immunostaining than control animals, which was also known from Kelatorphan- and particularly morphine-treated animals.

The expression of the c-Fos protein in the mammalian nervous system is closely related to the intracellular calcium (Ca^{2+}_i) concentration (Morgan and Curran 1989, 1991). Ca^{2+}-influx via voltage-dependent calcium channels, Ca^{2+}-channel activating drugs (Morgan and Curran 1986) or activation of excitatory amino acid (EAA) receptors (Lerea and McNamara 1993) may serve as the intracellular signal coupling synaptic activation with gene expression. (Fig. 5) Ionotropic glutamate receptor subtypes activate c-*fos* transcription, probably by distinct Ca^{2+} requiring intracellular signalling pathways (Lerea and McNamara 1993).

In the spinal cord the increase of Ca^{2+}_i in dorsal horn neurons is assumed to result from the release of EAAs, such as L-glutamate, and various peptide neurotransmitters, such as substance P (SP) from primary afferent fibers following peripheral noxious stimulation (Duggan et al. 1988). Since the amount of c-Fos protein is proportional to the degree of transsynaptic activation, the suppressive effect of carbamazepine, valproate and phenytoine on c-Fos expression is likely to emerge from the depression of neuronal discharge activity and the concomittant decrease in Ca^{2+}_i. The physiological mechanisms by which anticonvulsants can achieve this effect include (i) suppression of high-frequency repetitive firing by acting on the conductance of various ion channels (DeLorenzo 1986); (ii) postsynaptic potentiation of GABA responses, or a decrease of excitatory synaptic transmission and regulation of Ca^{2+}-dependent neurotransmitter release (DeLorenzo 1986); (iii) an augmented GABAergic inhibition by activation of blockage of enzymes involved in GABA-metabolism (Macdonald and McLean 1986). Phenytoine and carbamazepine reduced polysynaptic reflexes and post-tetanic potentiation of monosynaptic reflexes in the spinal cord (Krupp 1969) and increased the latency of synaptic responses in single neurons of the trigeminal nucleus *in vivo* (Fromm and Killian 1967). In spinal cord neurons *in vitro*, phenytoine, carbamazepine and valproate, at concentrations equivalent to therapeutic plasma levels, produced a voltage-dependent limitation of sustained repetitive firing without changing resting membrane properties and, at least in the majority of neurons, the postsynaptic responses to iontophoretically applied GABA and glutamate (Macdonald and McLean 1986). N-methyl-D-aspartate (NMDA)-activated currents were dose-dependently blocked by carbamazepine, but not by phenytoine (Lampe and Bigalke 1990). As c-Fos expression by noxious thermal stimulation (Tölle et al. 1991a; Wisden et al. 1990) was not affected by NMDA-antagonists, the suppressive effects of carbamazepine on c-Fos expression were probably not exerted via NMDA-receptor modulation.

Segmental and supraspinal GABA systems are both implicated in central pain modulation (Sawynok 1987). High levels of GABA and its synthesizing enzyme glutamate decarboxylase (GAD) were found in the superficial dorsal horn of normal rats (Hunt et al. 1981). GABA was increased in synaptic terminals (Löscher 1981) and various brain structures (Higuchi et al. 1986) by valproate-activated GAD (Löscher 1981) or by carbamazepine-inhibited GABA-degrading enzyme succinic semialdehyde dehydrogenase (Sawaya et al. 1975). Phenytoine had no effects on GABA levels in the brain (Patsalos and Lascelles 1981). Enhanced concentrations of GAD mRNA and GABA were found in the spinal cord following inflammation (Castro-Lopes et al. 1992, 1994a, 1994b). The involvement of GABA-mediated mechanisms in spinal antinociception (Sawynok 1987) is further

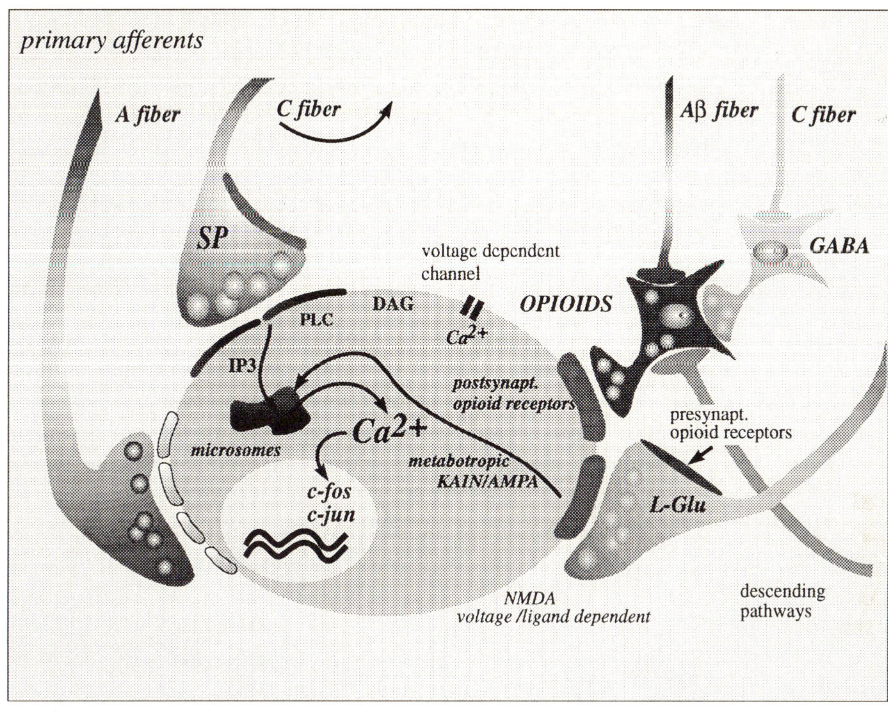

Fig. 5. Supposed mechanisms of c-Fos induction in dorsal horn neurons of the spinal cord. An increase in intracellular calcium (Ca^{2+}_i) is assumed to be the triggering event for c-Fos induction in neuronal tissue. Synaptic activation of dorsal horn neurons can increase Ca^{2+}_i through actions of various transmitters on different combinations of receptors.

Primary afferent fibers in the spinal cord appear to use SP and EAAs like L-Glu as transmitters. There is evidence that L-Glu and SP are co-localized in primary afferent fibers. Current research suggests that L-Glu exerts its function through at least five different receptors: NMDA, kainate, quisqualate depolarize most neurons by a conductance increase, the AP4 receptor may be an autoreceptor at a presynaptic site, and the recently identified ACPD receptor influences the IP3 metabolism through which also L-Glu might be able to increase Ca^{2+}_i and stimulate c-Fos expression. Direct evidence has been provided that SP can increase Ca^{2+}_i through voltage-gated channels and also through a metabotropic action via the IP3 system. The activation of the kainate and quisqualate receptors and the NMDA receptor triggers Ca^{2+}-influx through both voltage- and ligand-gated channels.

NMDA receptor mediated actions are blocked by the competitve antagonist APV and the non-competitve antagonist MK-801 and ketamine. Opioid peptide containing interneurons impinge on dorsal horn neurons and probably tonically inhibit these neurons, an effect which can be unmasked by the local application of naloxone. Systemic opioids exert their inhibitory effect on discharge activity of dorsal horn neurons probably through pre- and postsynaptically located receptors. Opioids (released from synaptic and non-synaptic sites) reduce the efficacy of topically or synaptically released L-Glu through an action mediated via postsynaptic opiate receptors. L-Glu and/or SP may also be reduced by a reduction voltage-gated Ca^{2+}-channels by the opiate agonists (presynaptic opiate receptors).

With regard to acute noxious thermal stimulation opiates - in contrast to NMDA antagonists - are effective in reducing c-Fos induction. It is suggested that the transient increase of Ca^{2+}_i evoked by NMDA is not essential to trigger the proto-oncogene expression. In other paradigms of chronic inflammatory pain NMDA receptor mediated components might prove to be more substantial. (Adapted from Tölle et al. 1991a)

al. (1993) investigated the effects of the specific α_2-adrenergic agonist medetomidine on the expression of c-Fos, Fos B, c-Jun, Jun B and Krox-24 proteins in the rat central nervous system. Administration of medetomidine dose-dependently diminished IEG expression in the spinal cord evoked by formalin injection into one hind paw. This reduction was also present in the parabrachial nucleus, and at a high dose in centromedian and paraventricular nuclei of the thalamus. Atipamezole (a specific α_2-adrenergic antagonist) attenuated the effect of metedomidine in the spinal cord and completely reversed its action in the parabrachial nucleus. However, in the medial thalamus, atipamezole increased the number of c-Fos positive neurons.

IEGs and target gene control

Although a causal relationship between peripheral stimulation, IEG induction and long-term changes in the expression of target genes is still lacking, circumstantial evidence indicates that at least some long-term effects, triggered by peripheral injury in central neurons, may be initiated by transactivation of IEGs. Compared to the numerous, partly co-induced (Naranjo et al. 1991), IEG labeled neurons observed following chronic noxious stimulation (Abbadie and Besson 1992a; Tölle et al. 1994c, 1994d), only a limited number of neurons, mostly in the substantia gelatinosa, changes mRNA levels for target genes, e.g. genes coding for opioid peptide precursors (Naranjo et al. 1991; Noguchi et al. 1992), or changes the amount of target gene protein as for e.g. dynorphin (Lanteri-Minet et al. 1993). This obvious discrepancy in the number of IEGs and target gene expressing neurons favours the idea that a highly developed overlap of various transcription factors has to be fulfilled in a single neuron to achieve transactivation of a specific secondary target gene or, alternatively, only genes with a low level of expression, such as Fos B, c-Jun or Jun D, have the capacity to induce long-term changes. The actual affinities, specificities and activities of these transcriptionally active complexes depend upon subunit composition, posttranslational modifications, competing ligands, flanking nucleotide sequences and overlapping binding sites (Morgan and Curran 1991; Rauscher et al. 1988; Ryseck and Bravo 1991). The level of complexity is extended even further by the finding that IEGs act as negative regulators of their own expression (Lazo et al. 1992; Sassone-Corsi et al. 1988). The functional relevance of these multi-combinatorial transcriptional complexes for DNA binding will depend upon the competition of the various ligands for DNA binding in the nucleus of a given cell. These complexes may uniquely regulate subsets of target genes containing a particular variant of the AP-1 site, or may have different effects upon the same set of regulatory elements (Sonnenberg 1989a, 1989b). The highest binding affinity to DNA is demonstrated for Fos B and c-Jun, followed by Jun D containing complexes (Ryseck and Bravo 1991). These IEGs were found in neurons located in regions associated with long-term changes in neuronal excitability in the spinal cord dorsal horn. The high transcriptional activity of several IEGs could thus trigger transcriptional operations and neuronal plasticity, even under the protection of a high dosage of morphine applied before noxious stimulation.

IEGs and central sensitization

Most of the currently available data on IEG expression by noxious stimulation are based on the IEG c-Fos and experiments employing acute or subacute stimuli. Conclusions, drawn from studies such as these, to explain mechanisms that may contribute to states of chronic pain in man (Coderre et al. 1993; Dubner and Ruda 1992; Fitzgerald 1990), should be treated with caution, as different mechanisms may operate in states of acute and chronic noxious stimulation. The experimental observations suggest that, provided the prevention of transcriptional operations is needed in order to avoid central sensitization, a more elaborate pharmacological treatment is required. The particular importance of the early phase of central hyperexcitation for initializing the stimulus-transcription cascade is yielded by pharmacological studies. In the spinal cord, morphine suppressed c-Fos induction only when applied *prior* to acute noxious stimulation but had no effect, even when injected immediately after noxious stimulation (Tölle et al. 1994b) (Fig. 4). Also in conditions of chronic disease, as in polyarthritic (Abbadie and Besson 1993) or mononeuropathic (Basbaum et al. 1991) rats, morphine or other analgesic compounds were not able to diminish the steady-state expression of c-Fos in the spinal cord, once the inoculation of the disease had started. Electrophysiological studies previously evidenced that morphine acts pre-emptively on the spinal cord (Dickenson and Sullivan 1986; Woolf and Wall 1986), and recent investigations with the autotomy model of neuropathic pain showed that self-mutilation was effectively prevented only when morphine was applied before unilateral sciatic nerve section (Puke and Wiesenfeld-Hallin 1993).

Moreover, clinical investigations provide considerable evidence that pre-emptive analgesia can prevent the induction of prolonged changes in the central neural function that may later contribute to postoperative pain (Katz et al. 1992; Wall 1988). Additionally, pretreatment with morphine and/or local anesthetics reduced the incidence of phantom pain, the scoring of postoperative pain and the request for analgesics (Bach et al. 1988; Kiss and Kilian 1992; McQuay et al. 1988). These clinical data underline the importance of pretreatment with opiates in the prevention of central sensitization including transcriptional operations (for review see Coderre et al. 1993).

Regarding anticonvulsants, our observations suggest that they are able to prevent long-term changes induced by activity-dependent gene modulation, not only in the hippocampus during seizures (White and Gall 1987). The maintained suppression of discharge activity in neurons, involved in pain signalling, by anticonvulsants could help to restore physiological discharge activity in these neurons by extinction of neuroplastic adaptive mechanisms, and help to alleviate pain and probably also spasticity.

References

Abbadie C, Besson JM (1992a) c-Fos expression in rat lumbar spinal cord during the development of adjuvant-induced arthritis. Neurosci 48:985-993

Abbadie C, Lombard MC, Morain F, Besson JM (1992b) Formalin injection in the forepaw: effects of dorsal rhizotomies. Brain Res 578:17-25

Abbadie C, Besson JM (1993) Effects of morphine and naloxone on basal and evoked FOS-like immunoreactivity in lumbar spinal cord neurons of arthritic rats. Pain 52:29-39

Abbadie C, Honore P, Fournie-Zaluski MC, Roques BP, Besson JM (1994a) Effects of opioids and non-opioids on c-Fos-like immunoreactivity induced in rat lumbar spinal cord neurons induced by noxious heat stimulation. Eur J Pharmacol 258:215-227

Abbadie C, Honore P, Besson JM (1994b) Intense cold noxious stimulation of the rat hinpaw induces c-Fos expression in lumbar spinal cord neurons. Neuroscience 59:457-468

Almeida A, Lima D, Coimbra A (1994) c-Fos activation of lamina I neurons projecting to the medullary dorsal reticular nucleus by different kinds of noxious stimulation. Eur J Neurosci Suppl 7:Abstr 21.08

Bach S, Noreng MF, Tj'ellchen NU (1988) Phantom limb pain in amputees during the first 12 months following limb amputation, after preoperative lumbar epidural blockade. Pain 33:297-301

Bartel DP, Sheng M, Lau LF, Greenberg ME (1989) Growth factors and membrane depolarization activate distinct programs of early response gene expression: dissociation of *fos* and *jun* induction. Genes Dev 3:304-313

Basbaum AI, Fields HL (1984) Endogenous pain control systems: Brainstem spinal pathways and endorphin circuitry. Annu Rev Neurosci 7:309-338

Basbaum AI, Presley R, Menetrey D, Chi SI, Levine JD (1989) Somatic, articular and visceral noxious-stimulus evoked expression of c-Fos in the spinal cord of the rat: differential patterns of activity and modulation by analgesic agents. In: Cervero F, Bennett GJ, Headley PM, (eds) Processing of sensory information in the superficial dorsal horn of the spinal cord. Vol 176, Plenum Press, New York, pp 365-381

Basbaum AI, Chi SI, Levine JD (1991) Factors that contribute to peripheral nerve injury-evoked persistent expression of the c-*fos* proto-oncogene in the spinal cord in the rat. In: Besson JM, Guilbaud G (eds) Lesions of primary afferent fibers as a tool for the study of clinical pain. Elsevier, Amsterdam, pp 205-218

Besse D, Lombard MC, Zajac JM, Roques BP, Besson JM (1990) Pre- and postsynaptic distribution of μ, δ and κ opioid receptors in the superficial layers of the cervical dorsal horn of the rat spinal cord. Brain Res 521:15-22

Besson JM, Chaouch A (1987) Peripheral and spinal mechanisms of nociception. Physiol Rev 67:67-154

Besson JM, Guilbaud G (1992) Towards the use of noradrenergic agonists for the treatment of pain. Excerpta Medica, Amsterdam

Birder LA, Roppolo JR, Iadarola MJ, deGroat WC (1991) Electrical stimulation of visceral afferent pathways in the pelvic nerve increases c-Fos in the rat lumbosacral spinal cord. Neurosci Lett 129:193-196

Birder LA, deGroat WC (1992a) Increased c-Fos expression in spinal neurons after irritation of the lower urinary tract in the rat. J Neurosci 12:4878-4889

Birder LA, de Groat WC (1992b) The effect of glutamate antagonists on c-Fos expression induced in spinal neurons by irritation of the lower urinary tract. Brain Res 580:115-120

Bullitt E (1989) Induction of c-fos-like protein within the lumbar spinal cord and thalamus of the rat following peripheral stimulation. Brain Res 493:391-397

Bullitt E (1990) Expression of c-fos-like protein as a marker for neuronal activity following noxious stimulation in the rat. J Comp Neurol 296:517-530

Bullitt E (1991) Somatotopy of spinal nociceptive processing. J Comp Neurol 312:279-290

Bullitt E, Lee CL, Light AR, Willcockson H (1992) The effect of stimulus duration on noxious-stimulus induced c-Fos expression in the rodent spinal cord. Brain Res 580:172-179

Butler SH, Godefroy F, Besson JM, Weil-Fugazza J (1992) A limited arthritic model for chronic pain studies in the rat. Pain 48:73-81

Castro-Lopes JM, Tavares I, Tölle TR, Coito A, Coimbra A (1992) Increase in GABAergic cells and GABA levels in the spinal cord in unilateral inflammation of the hindlimb in the rat. Eur J Neurosci 4:296-301

Castro-Lopes JM, Tavares I, Tölle TR, Coimbra A (1994a) Carrageenan-induced inflammation of the hind foot provokes a rise of GABA-immunoreactive cells in the rat spinal cord that is prevented by peripheral neurectomy or neonatal capsaicin treatment. Pain 56:193-201

Castro-Lopes JM, Tölle TR, Pan B, Zieglgänsberger W (1994b) Expression of GAD mRNA in spinal cord neurons of normal and monoarthritic rats. Mol Brain Res (in press)

Cervero F, Wolstencroft JH (1984) A positive feedback loop between spinal cord nociceptive pathways and antinociceptive areas of the cat's brainstem Pain 20:125-138

Chi SI, Levine JD, Basbaum AI (1993a) Peripheral and central contributions to the persistent expression of spinal cord Fos-like immunoreactivity produced by sciatic transection in the rat. Brain Res 617:225-237

Chi SI, Levine JD, Basbaum AI (1993b) Effects of injury discharge on the persistent expression of spinal cord Fos-like immunoreactivity produced by sciatic nerve transection in the rat. Brain Res 617:220-224

Chinapen S, Swann JM, Steinman JL, Komisaruk BR (1992) Expression of c-Fos protein in lumbosacral spinal cord in response to vaginocervical stimulation in rats. Neurosci Lett 145:93-96

Coderre TJ, Katz J, Vaccarino AL, Melzack R (1993) Contribution of central neuroplasticity to pathological pain: review of clinical and experimental evidence. Pain 52:259-285

Cole AJ, Saffen DW, Baraban JM, Worley PF (1989) Rapid increase of an immediate-early gene messenger RNA in hippocampal neurons by synaptic NMDA receptor activation. Nature (Lond) 340:474-476

Cruz F, Lima D, Avelino A, Coimbra A (1993) c-Fos activation of spinal neurons by chemical stimulation or distension of the urinary bladder. In: Abstracts of the 7th World Congress on Pain, IASP Press, pp 438-439

Cruz F, Lima D, Avelino A, Coimbra A (submitted) Noxious stimulation of the urinary bladder increases c-Fos expression at two spinal cord levels.

Dai JL, Zhu YH, Li KY, Quang DK, Xu SF (1993) Central Expression of c-Fos protein after peripheral noxious thermal stimulation in awake rats. Chung Kuo Yao Li Hsueh Pao 14:306-311

DeLeo JA, Coombs DW, McCarthy LE (1991) Differential c-Fos-like protein expression in mechanically versus chemically induced visceral nociception. Brain Res Mol Brain Res 11:167-170

DeLorenzo RJ (1986) A molecular approach to the calcium signal in brain: relationship to synaptic modulation and seizure discharge. In: Delgado-Escueta AV, Ward AA Jr, Woodbury DM, Porter RJ (eds) Advances in Neurology, Vol 44, Raven Press, New York, pp 435-464

Dickenson AH, Sullivan AF (1986) Electrophysiological studies on the effects of intrathecal morphine on nocicepitive neurons in the rat dorsal horn. Pain 24:211-222

Dragunow M, Robertson HA (1988) Localization and induction of c-Fos protein like immunoreactive material in the nuclei of adult mammalian neurons. Brain Res 440:252-260

Draisci G, Iadarola MJ (1989) Temporal analysis of increases in c-fos, preprodynorphin and preproenkephalin mRNAs in rat spinal cord. Mol Brain Res 6:31-37

Dubner R, Ruda MA (1992) Activity-dependent neuronal plasticity following tissue injury and inflammation. TINS 15:96-103

Duggan AW, Hendry IA, Morton CR, Hutchinson WD, Zhao ZQ (1988) Cutaneous stimuli releasing immunoreactive substance P in the dorsal horn of the cat. Brain Res 451:261-273

Evan G (1991) Regulation of gene expression. Br Med Bulletin 47:116-135

Fitzgerald M (1990) c-Fos and the changing face of pain. TINS 13:439-440

Fournié-Zaluski MC, Chaillet P, Bouboutou R, Coulaud A, Cherot P, Waksman G, Costentin J, Roques BP (1984) Analgesic effects of kelatorphan, a new highly potent inhibitor of multiple enkephalin degrading enzymes. Eur J Pharmacol 102:525-528

Fromm GH, Killian JM (1967) Effect of some anticonvulsant drugs on the spinal trigeminal nucleus. Neurology 17:275-280

Fuxe K, Agnati LF, Rosen L, Bjilke B, Cintra A, Bortolotti F, Tinner B, Andersson C, Hasselroth U, Steinbusch H, et al. (1991) Computer assisted image analysis techniques allow a characterization of the compartments within the basal ganglia. Focus on functional compartments produced by d-amphetamine activation of the c-*fos* gene and its relationship to the glucocorticoid receptor. J Chem Neuroanat 4:355-372

Giesler GJ, Yezierski RP, Gerhart KD, Willis WD (1981) Spinothalamic tract neurons that project to medial and/or lateral thalamic nuclei: evidence for a physiologically novel population of spinal cord neurons. J Neurophysiol 46:1285-1308

Gibson SJ, Polack JM, Bloom SR, Wall PD (1981) The distribution of nine peptides in rat spinal cord with special emphasis on the substantial gelatinosa and the area around the central canal (lamina X). J Comp Neurol 201:

Goelet P, Castellucci VF, Schacher S, Kandel ER (1986) The long and the short of long-term memory - a molecular framework. Nature 322:419-422

Gogas KR, Ahlgren SC, Levine JD, Basbaum AI (1990) Going for a walk induces widespread expression of the Fos protein product of the c-*fos* protooncogene in central motor systems. Soc Neurosc Abs Abstr:368.3

Gogas KR, Presley RW, Levine JD, Basbaum AI (1991) The antinociceptive action of supraspinal opioids results from an increase in descending inhibitory control: correlation of nociceptive behavior and c-Fos expression. Neuroscience 42:617-628

Hammond DL, Presley R, Gogas KR, Basbaum AI (1992) Morphine or U-50,488 suppresses Fos protein-like immunoreactivity in the spinal cord and nucleus tractus solitarii evoked by a noxious visceral stimulus in the rat. J Comp Neurol 315:244-253

Headley PM, Grillner S (1990) Excitatory amino acids and synaptic transmission: the evidence for a physiological function. Trends Pharmacol Sci 11:205-211

Herdegen T, Kovary K, Leah JD, Bravo R (1991a) Specific temporal and spatial distribution of Jun, Fos and Krox-24 proteins in the spinal neurons following noxious transsynaptic stimulation. J Comp Neurol 313:178-191

Herdegen T, Tölle TR, Bravo R, Zieglgänsberger W, Zimmermann M. (1991b) Sequential expression of Jun B, Jun D and Fos B proteins in rat spinal neurons: cascade of transcriptional operations during nociception. Neurosci Lett 129:221-224

Herdegen T, Walker T, Bravo R, Leah JD, Zimmermann M (1990) The Krox-24 protein, a new transcription regulating factor: expression in the rat nervous system following afferent somatosensory stimulation. Neurosci Lett 120:21-24

Higuchi T, Yamazaki O, Takazawa A, Kato N, Watanabe N, Minatogawa Y, Yamazaki J, Ohshima H, Nagaki S, Igarashi Y, Noguchi T (1986) Effects of carbamazepine and valproic acid on brain immunoreactive somatostatin and gamma-aminobutyric acid in amygdaloid-kindled rats. Eur J Pharmacol 125:169-175

Hirai SI, Ryseck RP, Mechta F, Bravo R, Yaniv M (1989) Characterization of *jun* D: a new member of the *jun* protooncogenes family. EMBO J 8:1433-1439

Hunt SP, Kelly JS, Emson PC, Kimmel JR, Miller RJ, Wu JY (1981) An immunohistochemical study of neuronal populations containing neuropeptides or gamma-aminobutyrate within the superficial layers of the rat dorsal horn. Neuroscience 6:1883-1898

Hunt SP, Pini A, Evan G (1987) Induction of c-Fos-like proteins in spinal cord neurons following sensory stimulation. Nature (Lond) 328:632-634

Jasmin L, Wang H, Tarczy-Hornoch K, Levine JD, Basbaum AI (1993) Noxious stimulus-evoked Fos-like immunoreactivity (FLI) in spinobrachial neurons in lamina I is not affected by morphine. In: Abstracts of the 7th World Congress on Pain, IASP Press, Abstr 606

Jones SL, Light AR (1990) Electrical stimulation in the medullary nucleus raphe magnus inhibits noxious heat-evoked Fos protein-like immunoreactivity in the rat lumbar spinal cord. Brain Res 530:335-338

Jones SL (1992) Noradrenergic modulation of noxious heat-evoked Fos-like immunoreactivity in the dorsal horn of the rat sacral spinal cord. J Comp Neurol 325:435-445

Katz J, Kavanagh BP, Sandler AN (1992) Does pre-operative fentanyl administration influence postoperative analgesia. Pain 51:123-124

Keay KA, Bandler R (1993) Deep and superficial noxious stimulation increases Fos-like immunoreactivity in different regions of the midbrain periaqueductal grey of the rat. Neurosci Lett 154:23-26

Kehl LJ, Gogas KR, Lichtblau L, Pollock CH, Mayes M, Basbaum AI, Wilcox GL (1991) The NMDA antagonist MK-801 reduces noxious stimulus-evoked fos expression in the spinal cord dorsal horn. In: Bond MR, Charlton JE, Woolf CJ (eds) Proceedings of the VIth Congress of Pain, Elsevier, Amsterdam, pp 307-311

Kiss IE, Kilian M (1992) Does opiate premedication influence postoperative analgesia? A prospective study, Pain 48:157-158

Krupp P (1969) The effect of Tegretal(R) on some elementary neuronal mechanisms. Headache 9:42-46

LaMotte C, Pert CB, Synder SH (1976) Opiate receptor binding in primate spinal cord: distribution and changes after dorsal root section. Brain Res 112:407-412

Lampe H, Bigalke H (1990) Carbamazepine blocks NMDA-activated currents in cultured spinal cord neurons. NeuroReport 1:26-28

Lanteri-Minet M, de Pommery J, Herdegen T, Weil-Fugazza J, Bravo R, Menetrey D (1993) Differential time course and spatial expression of Fos, Jun and Krox-24 proteins in spinal cord of rats undergoing subacute or chronic somatic inflammation. J Comp Neurol 333:223-235

Lazo PS, Dorfman K, Noguchi T, Mattci MG, Bravo R (1992) Structure and mapping of the fosB gene. Fos B downregulates the activity of the *fos* B promotor. Nucleic Acids Res 20:343-350

Leah JD, Sandkühler J, Herdegen T, Murashov A, Zimmermann M (1992) Potentiated expression of Fos protein in the rat spinal cord following bilateral noxious cutaneous stimulation. Neuroscience 48:525-532

Lerea LS, McNamara JO (1993) Ionotropic glutamate receptor subtypes activate c-*fos* transcription by distinct calcium-requiring intracellular signaling pathways. Neuron 10:31-41

Light AR, Kavookian AM (1988) Morphology and ultrastructure of physiologically identified substantia gelatinosa (lamina II) neurons with axons that terminate in deeper dorsal horn laminae (III-V). J Comp Neurol 267:172-189

Lima D, Avelino A, Coimbra A (1992) Differential participation of marginal cells in the transmission of nociceptive input to the thalamus and mesencephalon. Eur J Neurosci Suppl 5:91

Lima D, Avelino A, Coimbra A (1993) Differential activation of c-Fos in spinal neurones by distinct classes of noxious stimuli. NeuroReport 4:747-750

Lima D, Esteves F, Coimbra A (1994) c-Fos activation by noxious input of spinal neurons projecting to the nucleus of the tractus solitarius in the rat. In: Gebhart GF, Hammond DL, Jensen TS (eds) Proceedings of the 7th World Congress on Pain. Progress in Pain Research and Management, Vol 2, IASP Press, Seattle pp 423-434

Löscher W (1981) Valproate induced changes in GABA metabolism at the subcellular level. Biochem Pharmacol 30:1364-1366

Macdonald RL, McLean MJ (1986) Anticonvulsant drugs: mechanism of action. In: Delgado-Escueta AV, Ward AA, Woodbury DM, Wheal (eds) Basic Mechanisms of the Epilepsies: Molecular and Cellular Approaches. Adv Neurol, Vol 44 Raven Press, New York, pp 713-736

Matsumoto N, Kawarada K, Kamata K, Suzuki TA (1993) Electrical stimulation of tooth pulp increases the expression of c-Fos in cat supraoptic nucleus but not in the paraventricular nucleus. Life Sci 53:1235-1241

McQuay HJ, Carrol D, Moore RA (1988) Postoperative orthopaedic pain - the effect of opiate premedication and local anaesthetic blocks. Pain 33:291-295

Menetrey D, Gannon A, Levine JD, Basbaum AI (1989) The expression of c-Fos protein in presumed nociceptive interneurons and projection neurons of the rat spinal cord: anatomical mapping of the central effect of noxious somatic, articular and visceral stimulation. J Comp Neurol 285:177-195

Molander C, Xu Q, Grant G (1984) The cytoarchitectonic organization of the spinal cord in the rat. I. The lower thoracic and lumbosacral cord. J Comp Neurol 230:133-141

Morgan JI, Cohen DR, Hempstead JL, Curran T (1987) Mapping patterns of c-Fos expression in the central nervous system after seizure. Science 237:192-196

Morgan JI, Curran T (1986) Role of ion flux in the control of c-Fos expression. Nature 322:552-555

Morgan JI, Curran T (1989) Calcium as a modulator of the immediate-early gene cascade in neurons. Cell Calcium 9:303-311

Morgan JI, Curran T (1991) Stimulus-transcription coupling in the nervous system: involvement of the inducible proto-oncogenes *fos* and *jun*. Annu Rev Neurosci 14:421-51

Morton CR, Zhao ZQ, Duggan, AW (1987) Kelatorphan potentiates the effect of (Met5)enkephalin in the substantia gelatinosa of the cat spinal cord. Eur J Pharmacol 140:195-201

Mumenthaler M, (1976) The pathophysiology of pain. In: Birkmayer W (ed) Epileptic seizures-behaviour- pain, Huber, Bern, pp 275-312

Nagao M, Kamo H, Akiguchi I, Kimura J (1993) Induction of c-Fos-like protein in the lateral habenular nucleus by persistent noxious peripheral stimulation. Neurosci Lett 151:37-40

Naranjo JR, Mellström B, Achaval M, Lucas JJ, Del Rio J, Sassone-Corsi P (1991) Co-induction of *jun* B and c-*fos* in a subset of neurons in the spinal cord. Oncogene 6:223-227

Noguchi K, Dubner R, Ruda MA (1992) Preproenkephalin mRNA in spinal dorsal horn neurons is induced by peripheral inflammation and is co-localized with Fos and Fos-related proteins. Neuroscience 46:561-570

Patsalos PN, Lascelles PT (1981) Changes in regional brain levels of amino acid putative neurotransmitters after prolonged treatment with the anticonvulsant drugs diphenylhydantoin, phenobarbitone, sodium valproate, ethosuximide, and sulthiame in the rat. J Neurochem 36:688-695

Pertovaara A, Bravo R, Herdegen T (1993) Induction and suppression of immediate-early genes by selective alpha-2-adrenoceptor agonist and antagonist in the brain following noxious peripheral stimulation. Neuroscience 54:117-126

Presley RW, Menetrey D, Levine JD, Basbaum AI (1990) Systemic morphine suppresses noxious stimulus-evoked Fos protein-like immunoreactivity in the rat spinal cord. J Neurosci 10:323-335

Pretel S, Piekut DT (1991) Enkephalin, substance P, and serotonin axonal input to c-Fos-like immunoreactive neurons of the rat spinal cord. Peptides 12:1243-1250

Puke MJC, Wiesenfeld-Hallin Z (1993) The differential effects of morphine and alpha 2 adrenoceptor agonists clonidine and dexmedetomidine on the prevention and treatment of experimental neuropathic pain. Anesth Analg 77:104-109

Rauscher FJ III, Voulalas PJ, Franza BR Jr, Curran T (1988) Fos and Jun bind cooperatively to the AP-1 site: reconstitution *in vitro*. Genes Dev 2:1687-1699

Roques BP, Noble F, Dauge V, Fournié-Zaluski MC, Beaumont A (1993) Neutral endopeptidase 24.11: structure, inhibiton, and experimental and clinical pharmacology, Pharmacol Rev 87-146

Ryseck P, Bravo R (1991) c-Jun, Jun B, and Jun D differ in their binding affinities to AP-1 and CRE consensus sequences: effect of Fos proteins. Oncogene 6:533-542

Sassone-Corsi P, Sisson JC, Verma IM (1988) Transcriptional autoregulation of the proto-oncogene *fos*. Nature (Lond) 334:314-319

Sato M, Foong FW (1983) A mechanism of carbamazepine-analgesia as shown by bradykinin-induced trigeminal pain. Brain Res Bull 10:407-409

Sawaya MCB, Horton RW, Meldrum BS (1975) Effects of anticonvulsant drugs on the cerebral enzymes metabolizing GABA. Epilepsia 16:649-655

Sawynok J (1987) GABAergic mechanisms of analgesia: An update. Pharmacol Biochem Behav 26:463-474

Sharp FR, Griffith J, Gonzalez MF, Sagar M (1989) Trigeminal nerve section induces Fos-like immunoreactivity (FLI) in brainstem and decreases FLI in sensory cortex. Mol Brain Res 6:217-220

Sonnenberg JL, Macgregor-Leon PF, Curran T, Morgan JI (1989a) Dynamic alterations occur in the levels and composition of transcription factor AP-1 complexes after seizure. Neuron 3:359-365

Sonnenberg JL, Rauscher FJ III, Morgan JI, Curran T (1989b) Regulation of proenkephalin by *fos* and *jun*. Science 246:1622-1625

Strassman AM, Vos BP (1993) Somatotopic and laminar organization of Fos-like immunoreactivity in the medullary and upper cervical dorsal horn induced by noxious facial stimulation in the rat. J Comp Neurol 331:495-516

Swett JE, Woolf CJ (1985) The somatotopic organization of primary afferent terminals in the superficial laminae of the dorsal horn of the rat spinal cord. J Comp. Neurol 231:66-77

Tavares I, Lima D, Coimbra A (1993) Neurons in the superficial dorsal horn of the rat spinal cord projecting to the medullary ventrolateral tericular formation express c-Fos after noxious stimulation of the skin. Brain Res 623:278-286

Tölle TR, Ableitner A, Castro-Lopes JM, Zieglgänsberger W (1994d) c-Fos Protein, prodynorphin mRNA, and protein kinase C are altered with distinct spatial and temporal patterns in the spinal cord of monoarthritic rats. In: Gebhart GF, Hammond DL, Jensen TS (eds) Proc. of the 7th World Congress on Pain. Progress in Pain Res. and Manag., Vol 2, IASP Press, Seattle, pp 409-422

Tölle TR, Castro-Lopes JM, Coimbra A, Zieglgänsberger W (1990) Opiates modify induction of c-*fos* proto-oncogene in the spinal cord following noxious stimulation. Neurosci Lett 111:46-51

Tölle TR, Castro-Lopes JM, Evan G, Zieglgänsberger W (1991a) c-Fos induction in the spinal cord following noxious stimulation: prevention by opiates but not by NMDA antagonists. In: Bond MR, Charlton JE, Woolf CJ (eds) Pain Research and Clinical Management, Vol 4, Proc VIth World Congress on Pain, Elsevier, Amsterdam, pp 299-305

Tölle TR, Castro-Lopes JM, Schadrack J, Evan G, Zieglgänsberger W (in press) Anticonvulsants suppress c-Fos protein expression in the spinal cord neurons following noxious thermal stimulation. Exp Neurol

Tölle TR, Castro-Lopes JM, Zieglgänsberger W (1994c) Nociceptive stimulation induces immediate-early gene expression and increases GABA immunoreactivity in spinal neurons. In: Hökfelt T, Schaible HG, Schmidt RF (eds) Neuropeptides, Nociception and Pain, Chapman and Hall, London, Glasgow, Weinheim, New York, Tokyo, Melbourne, Madras, pp 385-402

Tölle TR, Evan G, Castro-Lopes JM, Schadrack J, Zieglgänsberger W (1991b) The effects of opiates and NMDA-antagonists on c-Fos and NGFI-A induction in the spinal cord reveals stimulus-specifity: comparison of noxious heat and acute inflammation. Eur J Neurosci Suppl 4:Abstr 4174

Tölle TR, Herdegen T, Schadrack J, Bravo R, Zimmermann M, Zieglgänsberger W (1994a) Application of morphine prior to noxious stimulation differentially modulates expression of Fos, Jun and Krox-24 proteins in rat spinal cord neurons. Neuroscience 54:305-321

Tölle TR, Schadrack J, Castro-Lopes JM, Evan G, Roques BP, Zieglgänsberger W (1994b) Effects of Kelatorphan and morphine before and after noxious stimulation on immediate-early gene expression in rat spinal cord neurons. Pain 56:103-112

Traub RJ, Herdegen T, Gebhart GF (1993) Differential expression of c-Fos and c-Jun in two regions of the rat spinal cord following noxious colorectal distension. Neurosci Lett 160:121-125

Waksman G, Hamel E, Fournié-Zaluski MC, Roques BP (1986) Autoradiographic comparison of the distribution of the neutral endopeptidase "enkephalinase" and of μ and δ opioid receptors in rat brain. Proc Natl Acad Sci USA 33:1523-1527

Wall PD (1988) The prevention of postoperative pain. Pain 33:289-290

White JD, Gall CM (1987) Differential regulation of neuropeptide and proto-oncogene mRNA content in the hippocampus following recurrent seizures. Mol Brain Res 3:21-29

Williams S, Evan G, Hunt SP (1990) Changing patterns of c-Fos induction in spinal neurons following thermal cutaneous stimulation in the rat. Neuroscience 36:73-81

Williams S, Pini A, Evan G, Hunt SP (1989) Molecular events in the spinal cord following sensory stimulation. In: Cervero F, Bennett GJ, Headley PM (eds) Processing of Sensory Information in the Superficial Dorsal Horn of the Spinal Cord, Plenum Press, New York, pp 273-282

Willis WD, Goggeshall RE (1991) Sensory Mechanisms of the Spinal Cord. Plenum Press, New York

Wisden W, Errington ML, Williams S, Dunnett SB, Waters c, Hitchcock S, Evan G, Bliss TVP, Hunt SP (1990) Differential expression of immediate-early genes in the hippocampus and spinal cord. Neuron 4:603 614

Woolf CJ, Wall PD (1986) Relative effectiveness of C primary afferent fibers of different origin in evoking a prolonged facilitation of the flexor reflex in the rat. J Neurosci 6:1433-1442

Yaksh TL (1981) Spinal opiate analgesia. Characteristics and principles of action. Pain 11:293-346

Yao GL, Tohyama M, Senba E (1992) Histamine-caused itch induces Fos-like immunoreactivity in dorsal horn neurons - effect of morphine pretreatment. Brain Res 599:333-337

Zerial M, Toschi L, Ryseck RP, Schuermann M, Müller R, Bravo R (1989) The product of a novel growth factor activated gene, *fos* B, interacts with JUN proteins enhancing their DNA binding activity. EMBO J 8:761-769

Zieglgänsberger W (1980) An enkepalinergic gating system involved in nociception? In: Costa E, Trabucchi M (eds) Neural Peptides and Neuronal Communication, Raven Press, New York, pp 425-430

Zieglgänsberger W (1986) Central control of nociception. In: Mountcastle VB, Bloom FE, Geiger SR (eds) Handbook of Physiology - The Nervous System IV, Williams & Wilkins, Baltimore, pp 581-656

Zieglgänsberger W, Tölle TR (1993) The pharmacology of pain signalling. Curr Opin Neurobiol 3:611-618

A novel face of immediate-early genes: transcriptional operations dominated by c-Jun and Jun D proteins in neurons following axotomy and during regenerative efforts

T. Herdegen, S. Brecht, C.E. Fiallos-Estrada, H. Wickert, F. Gillardon, S. Voss, R. Bravo[*], and M. Zimmermann
II. Institute of Physiology, University Heidelberg, Im Neuenheimer Feld 326, 69120 Heidelberg (Germany).
[*] Bristol-Myers Squibb Pharmaceutical Research Institute, Department of Molecular Biology, Princeton, NJ, 08543 (USA)

Dedicated to Prof. Dr. M. Zimmermann on the occasion of his 60th birthday.

Introduction

The investigation of immediate-early genes (IEGs) in neurons and glial cells represents a new research area in neurobiology. IEGs are rapidly expressed by trans-synaptic stimulation or transmembranous information transfer. A substantial number of IEGs encode for transcription factors (TF) which are involved in the control of gene transcription. The early appearance of TF offers a new access to elucidate the link between intentional stimuli and neuroplastic-reactive responses of neurons via alterations of protein synthesis. The expression of IEG encoded TF can be visualized by immunocytochemistry on the protein level (Herdegen et al. 1991a; Hunt et al. 1987; Kovary and Bravo 1991a,b; Morgan et al. 1987) or by *in situ* hybridization on the mRNA level (Cole et al. 1989; Wisden et al. 1990).

The main pool of data about IEG expression in the CNS has raised from studies about the expression of the c-*fos* gene and its protein product c-Fos (Hunt et al. 1987; Morgan et al. 1987; Sagar et al. 1988; reviewed by Herschman 1991; Morgan and Curran 1991). c-Fos is a representative of TF encoded by IEGs. In the cell nucleus, TF "trans"-act on specific regulatory DNA sites in the promotors and enhancers of target genes (Angel and Karin 1991; Bravo 1990; Sheng and Greenberg 1990). By binding to these DNA response elements, TF control the activation of large transcription complexes comprising the enzyme mRNA-polymerase II that exerts the elongation of primary RNA according to the DNA matrix. Thus, the expression of IEGs marks those particular neurons that respond to extracellular stimulation with alterations of gene expression and consequent protein synthesis.

Lasting *de novo* protein synthesis is required for another classical paradigm in neurobiology: regeneration and regenerative propensity of neurons following transection of nerve fibers. The reaction of neurons to axonal injury starts only after a latency of hours, but persists several months. The onset of this complex neuronal response, the cell body-reaction, comprises lasting morphological and molecular genetic changes during degeneration and the subsequent regenerative efforts.

Which are the transcriptional operating proteins regulating the specific and complex changes of effector protein synthesis following axotomy? Whereas several effector genes have been identified involved in the axonal elongation and synaptogenesis (reviewed by Grafstein 1986; Skene 1989; Strittmatter et al. 1992; Stürmer et al. 1992), the search for TF involved in the mammalian cell body-reaction was less successful.

In 1989 and 1990, we reported for the first time about the expression of the c-Jun transcription factor protein in axotomized neurons (Herdegen et al. 1989, 1990) which is encoded by the IEG and proto-oncogene c-*jun* (Angel et al. 1988; Bravo 1990; Ryder et al. 1988; Ryseck et al. 1988; reviewed by Bravo 1990; Vogt and Bos 1990). In the course of these research activities, the evidence arose that the temporo-spatial expression patterns of structurally and functionally related IEG proteins substantially differ between axotomy and transsynaptic stimulation: transsynaptic and transmembranous information transfer evokes rather transient transcriptional operations dominated by c-Fos and Jun B (Gass et al. 1992a,b; Herdegen et al. 1991a,d, 1993e; Pertovaara et al. 1993), whereas axotomy evokes transcriptional operations dominated by c-Jun and Jun D (Herdegen et al. 1991b, 1992a, 1993a,b; Leah et al. 1991) without participation of c-Fos and Jun B proteins.

In the present volume, we summarize our findings about the expression of transcription factor proteins in axotomized neurons of peripheral and central nervous system and in cell culture system. We address to the relationship between axotomy and c-Jun expression, and to c-Jun as marker for the regenerative propensity. Finally, we discuss the (axonal) signal transfer and intracellular mechanisms selectively inducing those IEGs that might trigger the alterations in protein synthesis underlying the cell body-reaction.

Material and methods

Immunocytochemistry

The expression of all proteins was investigated by immunocytochemistry, and the antigene-antibody reactions were visualized by the avidin-biotin-complex method (Vectastain, Vector Lab., USA). The stainings were performed on 20 µm sections fixed on gelatine-covered slides (peripheral nerve fibers, dorsal root ganglia, ganglion nodosum, retinal ganglion cell layer) and on free floating 35-50 µm sections (spinal cord, brain); all tissues were cyroprotected in 30% sucrose and cut by cryostat. For fixation, the deeply anesthetized animals were perfused and postfixed for 24 h with 4% paraformaldehyde.

Transection of nerve fibers

All experiments were performed in deeply anaesthetized male Sprague-Dawley rats (pentobarbital 60 mg/kg b.w., i.p.). The sciatic nerve was transected at the sciatic notch, the vagus nerve was transected 5-10 mm inferior to the nodose ganglion, the cervical sympathetic trunk was resected over a length of 5-10 mm close to its entrance into the caudal pole of the superior cervical ganglion (Herdegen et al. 1991b).

The medial forebrain bundle (MFB) and the mammillo-thalamic tract (MT) were stereotaxically transected (Bregma - 2.3 frontal and 1.5 lateral) (Herdegen et al. 1993b). Transection of fimbria-fornix was performed at Bregma -2.8 frontal, and between 2.5 and 4.5 lateral (Brecht et al. in preparation). Axotomy of granular neurons of the cerebellum was performed according to the protocol of Lehmann et al. 1993.

Optic nerve was intraorbitally crushed in female Lewis rats and intracranially cut in the goldfish (Herdegen et al. 1993a).

Results

Summary: expression of c-Jun and Jun D are novel constituents of the neuronal cell-body response due to axotomy.

The expression of Jun, Fos, Krox and CREB transcription factor proteins was investigated in various axotomized neuronal populations following transection of peripheral and central nerve fibers (summarized in Table 1).

The lasting presence of c-Jun and Jun D is a constant finding in axotomized neurons in peripheral, central extrinsic neurons (projection of axons into the peripheral nervous system, PNS) and central intrinsic neurons (projection within the central nervous system, CNS) (Fiallos-Estrada et al. 1993; Herdegen et al. 1991b, 1992a, 1993a-c; Leah et al. 1991, 1993). A less pronounced concomitant increase and persistance of Krox-24 is also a reproducible finding in axotomized central intrinsic but not central extrinsic neurons (Herdegen et al. 1992a, 1993a-b). The immunoreactivities of the IEG encoded proteins Jun B, c-Fos, Fos B and Krox-20 do not change following axotomy in PNS and CNS neurons (Herdegen et al. 1991b, 1992a, 1993a,b,d; Leah et al. 1991, 1993).

A major focus of our studies also addresses to the expression of constitutively expressed transcription factors such as CREB (Calcium/cAMP response-element binding protein). Activation of CREB by phosphorylation is one prerequisite for the induction of IEGs (Lamph et al. 1990; Montminy et al. 1990; Sheng et al. 1990). Interestingly, transection of nerve fibers evokes a *decrease* of CREB-IR in a subset of axotomized neurons (Herdegen et al. 1992a,b, 1993b). Table 2 summarizes the effect of axotomy on the expression of transcription factors of the Jun, Fos, Krox and CREB protein families.

Type of transected nerve fibers	Population of axotomized neurons
A. Peripheral and cranial nerve fibers	
sciatic nerve	L4-L6 dorsal root ganglia
	L4-L6 motoneurons of spinal cord
saphenous nerve	L2-L3 dorsal root ganglia
cervical sympathetic trunk	pregangl. sympathetic neurons
vagus nerve	nodose ganglion
	dorsal motonucleus of vagus nerve
	ncl. ambiguus
facial nerve[*,**]	motonucleus of facial nerve
hypoglossus nerve[**]	motonucleus of hypoglossus nerve
B. central nerve fibers	
dorsal columns	lamina IV-V of spinal cord
medial forebrain bundle	substantia nigra compacta
	ventral tegmental area
	parafascicular ncl. of thalamus
mammillo-thalamic tract	ncl. mammillaris
optic nerve	retinal ganglion cell layer
fimbria-fornix	medial septal nucleus
cerebellum	granular layer of cerebellum

Table 1. Expression of c-Jun protein in axotomized neurons of the adult rat: a novel constituent in the neuronal cell body-response. The table summarizes the types of transected nerve fibers and the population of axotomized neurons that express c-Jun protein following axotomy. [*,**] Transection and perfusion of the rats was performed by C. Haas[*] (Haas et al. 1993) and by Neiss et al.[**] (Neiss et al. 1992).

Transection of peripheral nerve fibers

The onset of c-Jun and Jun D expression starts between 10 h and 24 h following transection of peripheral nerve fibers (sciatic, saphenous and vagus nerve) in a distance-dependent manner, being first visible in primary afferent neurons and later on in motoneurons (Fig. 1A,B) (Herdegen et al. 1991b, 1992a; Leah et al. 1991). Following crush of sciatic and vagus nerve with consequent re-innervation of the target, these Jun immunoreactivities return to basal levels between 30 and 50 days (Herdegen et al. 1991b; Leah et al. 1991). If the axonal elongation is prevented by ligation of the proximal stump, the increased c-Jun-IR persists up to 15 months in predominantly small diameter DRG neurons and up to 3 months in motoneurons (Herdegen et al. 1992a, 1993c). Selective expression of c-Jun and Jun D were also observed in primary afferent neurons and motoneurons of vagus nerve following vagotomy, in facial and hypoglossal motoneurons following transection of facial and hypoglossal nerve fibers and in presynaptic sympathetic neurons following transection of the sympathetic cervical trunk (Fig. 1C-H) (Herdegen et al. 1991b; Haas and Herdegen as well as Neiss and Herdegen, unpublished observations). These findings are summarized in Tab. 1,2.

A second transection of sciatic nerve, 100 days after the first transection and ligation, evoked a dramatic up-regulation of c-Jun in axotomized neurons which surpassed the intensity of c-Jun-IR visible after the first transection (Fig. 2) (Wickert et al. in preparation). This re-induction of c-Jun parallels the findings about enhanced regenerative efforts of repeatedly damaged neurons (Forman et al. 1980) and demonstrates that the slow decrease of c-Jun following the first lesion is not due to neuronal cell death or severe damage of the neuronal protein synthesis.

Table 2. Selective expression of transcription factor proteins in axotomized neurons of the rat

axotomized neurons	c-Jun	JunB	JunD	c-Fos	FosB	K-20	K-24	CREB
sciatic DRG	↑↑	-	↑↑	-	-	-	-	↑/-
sciatic motoneurons	↑↑	-	↑↑	-	-	-	-	↓
saphenous DRG	↑↑	-	↑↑	-	-		-	
facial ncl.	↑↑	-	↑↑	-	-		-	↓
hypoglossal ncl.	↑↑	-	↑↑	-	-		-	↓
dorsal vagal ncl.	↑↑	-	↑↑	-	-		-	↓
ncl. ambiguus	↑↑	-	↑↑	-	-			
preggl. symp. neurons	↑↑	-	↑↑	-				
retinal gangl. cells	↑↑	-	↑↑	-	-		↑	↓
mammillary ncl.	↑↑	-	↑↑	-	-	-	↑	-
substantia nigra	↑↑	-	↑↑	-	-	-	↑	↓
ventral tegmentum	↑↑	-	↑↑	-	-	-	↑	-
parafascicular ncl.	↑↑	-	↑↑	-	-	-	↑	-
medial septal ncl.	↑↑	-	↑↑	-	-	-	↑	
cerebellum	↑↑	-	↑↑	-	-		↑	

↑↑ = strong increase, ↑ = medium increase, - no change and absence, respectively, ↓ = decrease of immunorectivity (intensity of labelling and number of labelled neurons). The remaining proteins and axotomized neurons, respectively, were not investigated.

Transection of central nerve fibers
Transection of the medial forebrain bundle (MFB) and the mammillo-thalamic tract (MT).
Following stereotaxic transection of the MFB and MT (Herdegen et al. 1993b), c-Jun, Jun D and Krox-24 are uniformly induced between 24 and 36 h in the axotomized neuronal populations such as substantia nigra compacta (SNC), ventral tegmentum, ncl. parafascicularis of thalamus and mammillary body (Fig. 3A,C; Tab. 1,2). In contrast to the uniform onset of expression, the persistance of IEGs reveals dramatic differences: Jun and Krox decline to basal levels between 10 and

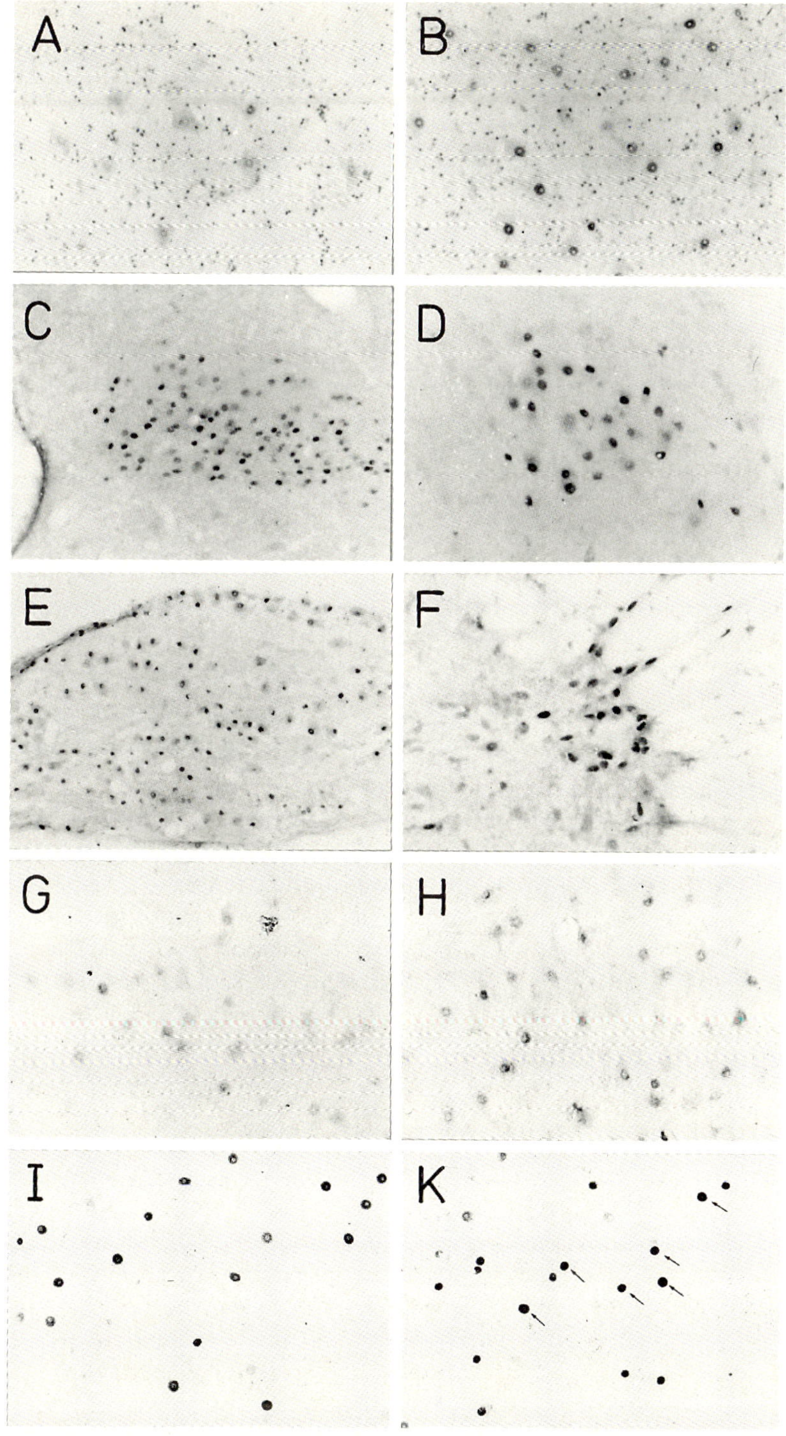

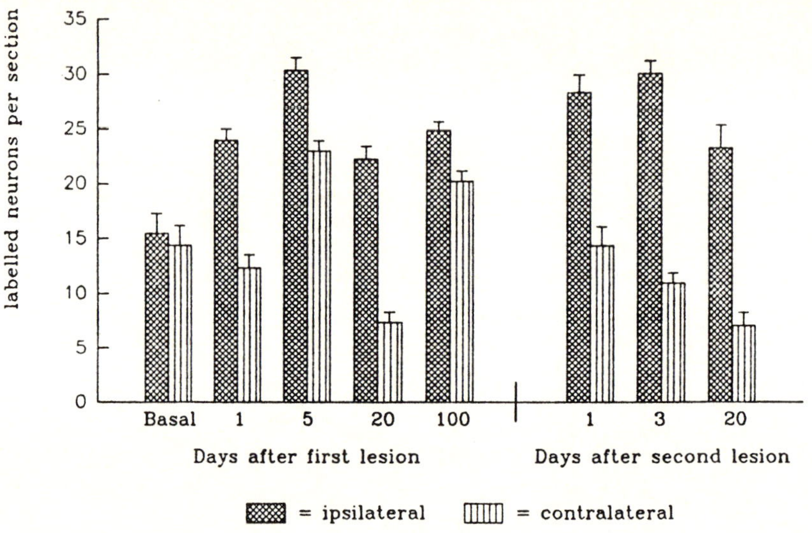

Fig. 2. Re-induction of c-Jun expression in sciatic motoneurons following a second transection of sciatic nerve 100 days after the first transection. This finding demonstrates that motoneurons can re-express the c-Jun protein even 100 days following a peripheral nerve cut.

20 days in substantia nigra and ventral tegmentum, respectively. In the parafascicular thalamic nucleus, however, the decrease does not start before 75 days, and axotomized neurons of the mammillary body express c-Jun, Jun D and Krox-24 on an elevated level at last for 150 days, the end of the observation period (Fig. 3B,D). Transient expression of c-Jun was also observed followed chemical destruction of rat nigral neurons by 6-hydroxydopamine (Jenkins et al. 1993a).

The early decrease of TF occurs in the dopaminergic neurons of substantia nigra compacta and ventral tegmentum which derive from common progenitor cells. Dopaminergic neurons have a restricted capacity for sprouting and survival *in vitro* (Fawcett 1992; Björklund et al. 1971); thus, our findings about the early decrease of c-Jun and Jun D expression following axotomy point out a relationship between the transient expression of TF and the reported restricted capacity for regenerative propensity.

Jun B, c-Fos and Fos B were not expressed in axotomized neurons; however, an early and transient expression of all Jun, Fos and Krox proteins was visible in

← **Fig. 1.** Jun D in sciatic motoneurons of (A) untreated rats and (B) 10 days following sciatic nerve cut. In contrast to motoneurons, expression of JunD does not change in glial cells. Increase of c-Jun-IR in axotomized neurons of (C) dorsal vagal ncl., (D) ncl. ambiguus and (E) nodose ganglion 10 days following vagotomy. (F) Increase in c-Jun-IR in preganglionic sympathetic neurons 20 days following transection of the cervical sympathetic trunk. c-Jun in facial motoneurons of (G) untreated rats and (H) its increase 5 days following transection of facial nerve (Haas et al. 1993; Haas and Herdegen, unpublished observation). Sciatic nerve transection (K) close to the dorsal root ganglia evokes in axotomized motoneurons an increase in intensity of c-Jun-IR (arrows) compared to (I) transection at the level of the sciatic notch.

substantia nigra reticularis, probably by transsynaptic impulse discharge. The IEG expression in axotomized and non-axotomized neurons could be differentiated by double-labelling of IEG and tyrosine-hydroxylase (TH), the latter expressed in dopaminergic neurons of substantia nigra compacta and ventral tegmentum. The early and transient IEG expression was present exclusively in TH-negative neurons, whereas the delayed and more persistent c-Jun expression occurred in TH-immunoreactive neurons (Herdegen et al. 1993b).

Transection of the fimbria-fornix bundle (FF).
Transection of FF axotomizes neurons in the medial septum. Following axotomy, these neurons show within 12 h an expression of c-Jun and Jun D. c-Jun persists at least for 100 days whereas increased Jun D-IR declines to basal levels after 60 days (Brecht et al. in preparation). Krox-24-IR was only slightly increased in neurons ipsilateral to FF transection. Expression of Jun B, c-Fos and Fos B could not be detected (Fig. 3E).

Axotomy of cerebellar neurons.
Selective expression of c-Jun, Jun D and Krox-24 is also visible in axotomized neurons of cerebellar granular layer (Fig. 3F) (Brecht et al. in preparation). Within 3 h, a coincident and transient expression of c-Fos, c-Jun, Jun B, Jun D and Krox-24 was visible in the damaged area probably by transsynaptic stimulation due to the mechanical injury. However, 24 h later, only c-Jun, Jun D and Krox-24 were still visible in the axotomized compartment and these IR persisted at least for 20 days.

Transection of the optic nerve.
Intraorbital crush of rat optic nerve is followed by a fairly rapid expression of c-Jun, Jun D and Krox-24 in the axotomized retinal ganglion cells (RGC) (Herdegen et al. 1993a). Within 24 h, numerous RGC are already labelled and the maximal number of labelled nuclei is seen after 5 days (Fig. 3G). After 8 days, when RGC have stopped their initial axonal sprouting (so-called abortive sprouting), all the immunoreactivities have already declined. This decrease also precedes the dramatic neuronal RGC death and the reduction of protein synthesis starting at the end of the first week (Villegas-Perez et al. 1992).

In contrast to mammalians, fishes elongate and regenerate their transected intrinsic central nerve fibers and form functional synaptic contacts (Stürmer et al. 1992). Transection of goldfish optic nerve evoked in axotomized RGC within 24 h the appearance of a Jun-IR the specific components of which could not be determined by specific antisera (Herdegen et al. 1993a). This Jun-IR persisted on a maximal plateau up to 30 days (Fig. 3H) when the RGC axons have reached their target sites in the tectum and start synaptogenesis. Between 50 and 100 days, Jun-IR declined to basal levels; during this time the elevated levels of protein synthesis and the increased amount of axonally transported proteins have returned to their basal levels (McKerracher and Hirscheimer 1992).

Again, c-Fos, Fos B and Jun B proteins were not expressed in RGC neither in the rat nor in the goldfish.

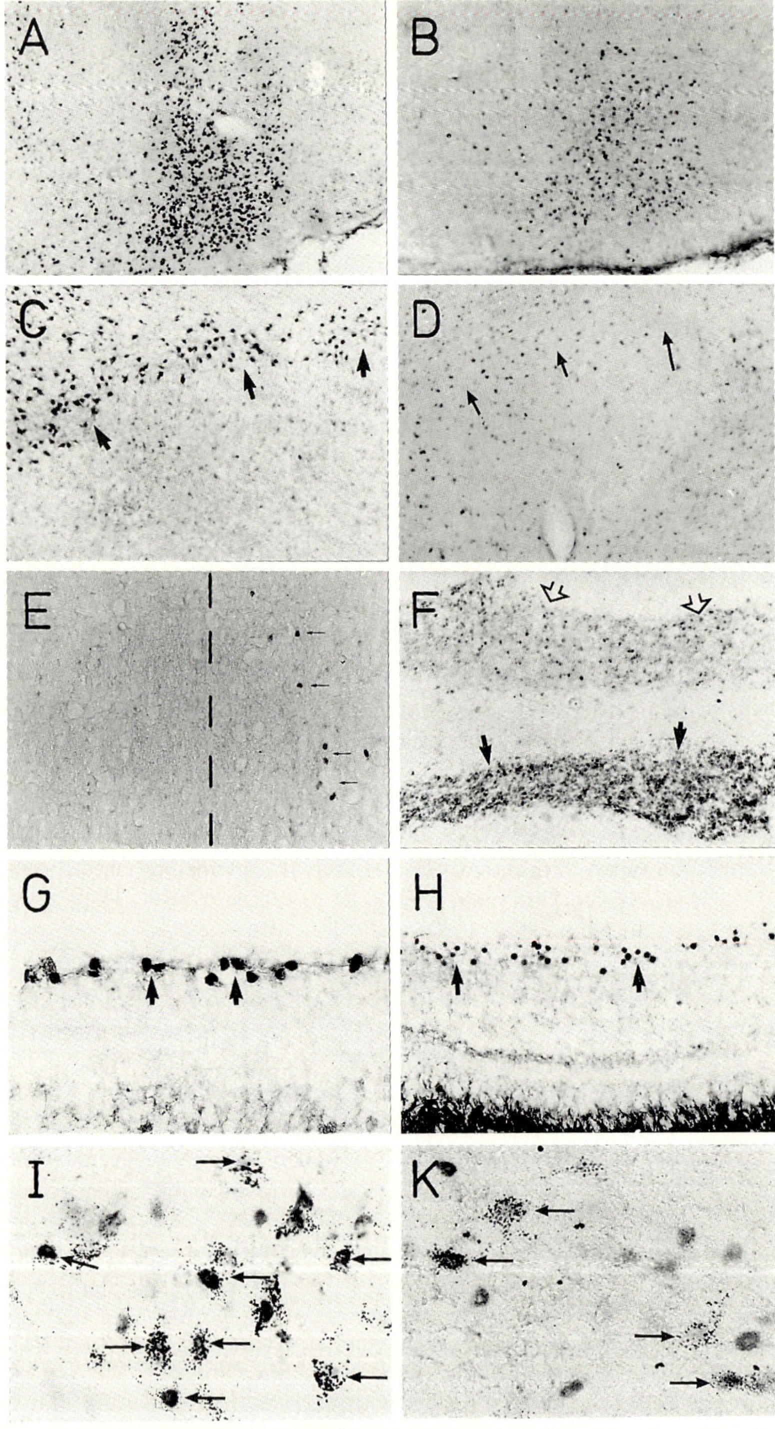

Changes of CREB expression

Axotomy also evokes alterations of the constitutively expressed transcription factor CREB (Tab. 2). CREB holds a superior position in the transcription hierarchy controlling the expression of both, the immediate-early genes e.g. *c-jun*, *jun* B, *c-fos* and *krox*-24, and the effector genes. We found a distinct decrease of CREB-IR in axotomized sciatic (Herdegen et al. 1992a) and facial motoneurons (Haas and Herdegen; Neiss and Herdegen, unpublished observations) as well as in axotomized RGC of both rat and goldfish (Herdegen et al. 1993a). The decrease of CREB-IR from its constitutive levels starts when c-Jun-IR has reached its maximal expression, and the CREB-IR reaches again its basal level after 2-3 months when c-Jun had declined to its low basal expression. It remains to be elucidated whether these inverse changes of c-Jun and CREB-IR reflect a mutual control of c-jun and CREB expression or just reflect the onset and termination of transcriptional operations underlying the axotomy-induced alterations of protein synthesis.

Ten days following MFB transection, we observed a decrease of CREB in neurons labelled by TH, which marks the axotomized neurons of SNC, whereby these TH neurons co-express c-Jun. This observation clearly demonstrates that up-regulation of c-Jun and suppression of CREB occurs in the very same neuron subsequently to axotomy, and that c-Jun can be expressed in absence of CREB (Brecht et al. 1994).

c-Jun is expressed in axotomized neurons

We have demonstrated by retrograde tracing that c-Jun is virtually expressed in the axotomized neurons. First, following sciatic nerve cut the fluorescence tracer fast blue was applied onto the proximal sciatic nerve stump; after 5 days, an almost perfect colocalization of c-Jun-IR in fast blue labelled neurons was assessed in the injured primary sensory neurons and motoneurons of sciatic nerve (Leah et al. 1991). Second, HRP-labelled gold was injected into the dorsolateral striatum that belongs to the areas of termination of substantia nigra compacta neurons; 4 days later, when the HRP-labelled gold has been already retrogradly

← **Fig. 3.** Differential persistance of c-Jun-IR in axotomized central intrinsic neurons (A-D): maximal c-Jun-IR in (A) mammillary body and (C) substantia nigra compacta (arrows) 5 days following transection of MFB and MT. However, c-Jun-IR (B) persists up to 150 days in mammillary body, but (D) is no longer increased in substantia nigra compacta (arrows) 20 days post-axotomy. (E) c-Jun-IR in axotomized neurons of medial septal ncl. (arrows) 20 days following fimbria-fornix transection; the dotted line indicates the midline. (F) c-Jun in granule cells (closed arrows) 5 days following transection of cerebellar granule cell layer; the open arrows indicate the basal c-Jun-IR of intact granule cells. (G) c-Jun in axotomized retinal ganglion cells (arrows) in the rat 5 days following intraorbital optic nerve crush. (H) Jun-IR in axotomized retinal ganglion cells (arrows) in the goldfish 20 days following intracranial optic nerve cut. (I) c-Jun and (K) Jun D labelled nuclei in those neurons of substantia nigra compacta 4 days following MFB transection that are also labelled by HRP-coupled gold (arrows). HRP-coupled gold was injected in the dorsolateral caudate putamen, one major area of nigral ascendent projections, 5 days prior to MFB transection.

transported to the cell body of the nigral projection neurons, the medial forebrain bundle was transected. After another 4 days, the rats were killed; almost all neurons labelled by the HRP-labelled gold also expressed the c-Jun and Jun D protein (Fig. 3I-K) (in collaboration with F. Anton).

Axotomy evokes long-lasting expression of c-*jun* mRNA

We demonstrated by polymerase chain reaction (PCR) that the persistance of c-Jun protein is due to lasting expression of c-*jun* mRNA. Ipsilateral and contralateral L3-L5 dorsal root ganglia (DRG) were homogenized 1 and 5 days following sciatic nerve cut. Following reverse-transcriptase/PCR, the amplification of c-*jun* mRNA in ipsilateral DRG evoked a distinctly higher signal compared to the c-*jun* mRNA of the contralateral DRG (Fig. 4). Because c-Jun-IR is not increased in glial cells (Herdegen et al. 1992a,b), the increased level of c-*jun* mRNA detected by RT-PCR demonstrates the up-regulation of c-*jun* mRNA in neurons following axotomy.

Expression of c-jun in
DRG following axotomy

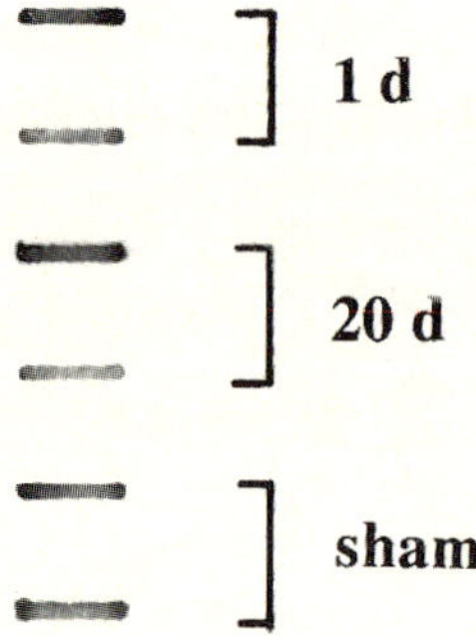

Fig. 4. Expression of c-*jun* mRNA in L4-L5 dorsal root ganglia (DRG) 1 d and 20 d following transection of sciatic nerve and in sham operated controls. Each of the upper bands refers to the ipsilateral, each of the lower bands refers to the contralateral DRG. Fluorescein-11-dUTP labelled primers complementary to specific c-*jun* sequences were amplified by AMV-reverse transcriptase and thermus thermophilus polymerase (Tth Pol), and the dimerization products were visualized by enhanced chemiluminescence (ECL). Our findings show an increased c-jun expression of axotomized sciatic DRG between 1 d and 20 d compared to the contralateral intact DRG; sham operation (sham) does not enhance c-*jun* mRNA levels.

c-Jun and colocalization of effector gene encoded proteins: nitric oxide synthase, tyrosine-hydroxylase, galanin and CGRP

The investigation of co-expression of TF and effector gene encoded proteins follows two purposes: first, it offers a valuable tool for screening of those target genes co-controlled by inducible transcription factors such as Jun, Fos and Krox protcins. Second, the expression patterns of both transcription factors and effector gcnes give further insight into functional processes following pathophysiological events such as during the degeneration and/or regeneration processes.

c-Jun and nitric oxide synthase (NOS) in axotomized neurons of PNS and CNS. The neuronal enzyme NOS catalyzes the transformation of arginin into citrulline thereby producing the gas radical nitric oxide (NO). NO belongs to a new group of molecules fulfilling the properties of neurotransmitters (reviewed by Bredt and Synder 1992; Moncada et al. 1991). The enzymatic activity of NOS can be visualized by the NADPH-diaphorase reaction (NDP). Previously, it has been demonstrated that NDP-labelled neurons are selectively resistant against ischemic, toxic and degenerative insults (reviewed by Bredt and Synder 1992; Moncada et al. 1991). Therefore, we have been interested in the question whether expression of NOS is altered by axotomy and whether these changes are related to the expression of TF such as c-Jun.

Axotomy alters the expession of NOS - but these alterations are not uniform: expression of NOS is induced, suppressed and remains unchanged in dependence of the type of axotomized neuron and, probably, on the intensity of neuronal damage (Tab. 3).

Following sciatic nerve cut, NOS-IR and NDP increase with a delay between 5 and 10 days in DRG, and decrease between 30 and 50 days (Fig. 5A,B) (Fiallos-Estrada et al. 1993); the increase of NOS-IR was due to mRNA expression (Verge et al. 1992). Axotomized sciatic motoneurons show neither change of their weak basal NOS-IR nor of the absent NDP (Fiallos-Estrada et al. 1993) (Tab. 3). However, an increase of NDP was reported in lumbar motoneurons following aversion representing a more severe damage of motoneurons with subsequent death of affected motoneurons (Wu 1993).

Table 3. Changes of nitric oxide synthase (NOS) immunoreactivity and NADPH-diaphorase reaction in axotomized neurons in the rat nervous system

sciatic prim. afferent neurons	↑	mammillary body	↑
sciatic motoneurons	o	substantia nigra	↓
hypoglossal motoneurons	o	parafascicular ncl.	o
facial motoneurons	o	granular layer of cerebellum	↓
		medial septal ncl.	↓

↑ = increase, ↓ = decrease, o = no change and absence, respectively, of NOS-immunoreactivity and NADPH-diaphorase reaction in c-Jun labelled neurons.

In the brain, a long-lasting increase of NOS expression with delayed onset is induced in axotomized neurons of the mammillary body which also have a basal NOS-IR (Fig. 5C,D). In contrast, neurons of substantia nigra, ventral tegmentum and ncl. parafascicularis of thalamus do not respond to axotomy with expression of NOS (Herdegen et al. 1993b). In axotomized neurons of the the cerebellum and the medial septum, NOS-IR distinctly decreases from a high basal level (Fig. 5E,H). NDP parallels these changes of NOS-IR. The up- and down-regulation of NOS-IR is mainly restricted to those neurons which co-express the c-Jun protein (Brecht et al. in preparation) (Tab. 3).

c-Jun and tyrosine-hydroxylase (TH) in substantia nigra compacta.
In axotomized neurons of substantia nigra compacta, expression of c-Jun is highly colocalized with TH positive neurons (Brecht et al. 1994). TH-IR declines following a transient increase due to accumulation after 3 days. Between 10 and 60 days, when c-Jun and TH are only visible in few scattered neurons of substantia nigra compacta, there is still a high degree of colocalization (Brecht et al. 1994). This observation suggests that c-Jun marks those neurons with a residual production of protein synthesis.

c-Jun, galanin and calcitonin gene-related peptide (CGRP).
Following sciatic nerve cut, appearance of c-Jun precedes the induction of both galanin in primary sensory neurons and CGRP in motoneurons; these neuropeptides belong to the class of rapidly induced effector gene encoded proteins. Up to 50 days, the colocalization values between c-Jun and the neuropeptides remained on a high level and above random probability, being true for the maximum of expression as well as for the return to suprabasal levels. The synthesis of CGRP underlies the re-formation of acetylcholine receptor in muscle fibers (New and Mudge 1986); and galanin inhibits the excitation of deafferented spinal neurons (Xu et al. 1990) thereby counteracting the hyperalgesia following peripheral nerve fiber lesion.

Mechanisms of c-Jun induction

The evaluation of the signal(s) responsible for the production of transcription factors most likely uncovers those intraneuronal processes underlying the changes of axotomy-related protein synthesis.

Distance between site of lesion and affected cell body.
The closer the site of lesion to the cell body the stronger the expression of c-Jun (Fig. 1I-K,6) (Leah et al. 1991). The high expression of c-Jun following axotomy close to the perikaryon and following removal of the proximal nerve stump excludes the possibility of c-Jun induction by neurotrophic factors.

Block of axonal transport.
Block of the axonal transport by application of colchicine on the *intact* sciatic

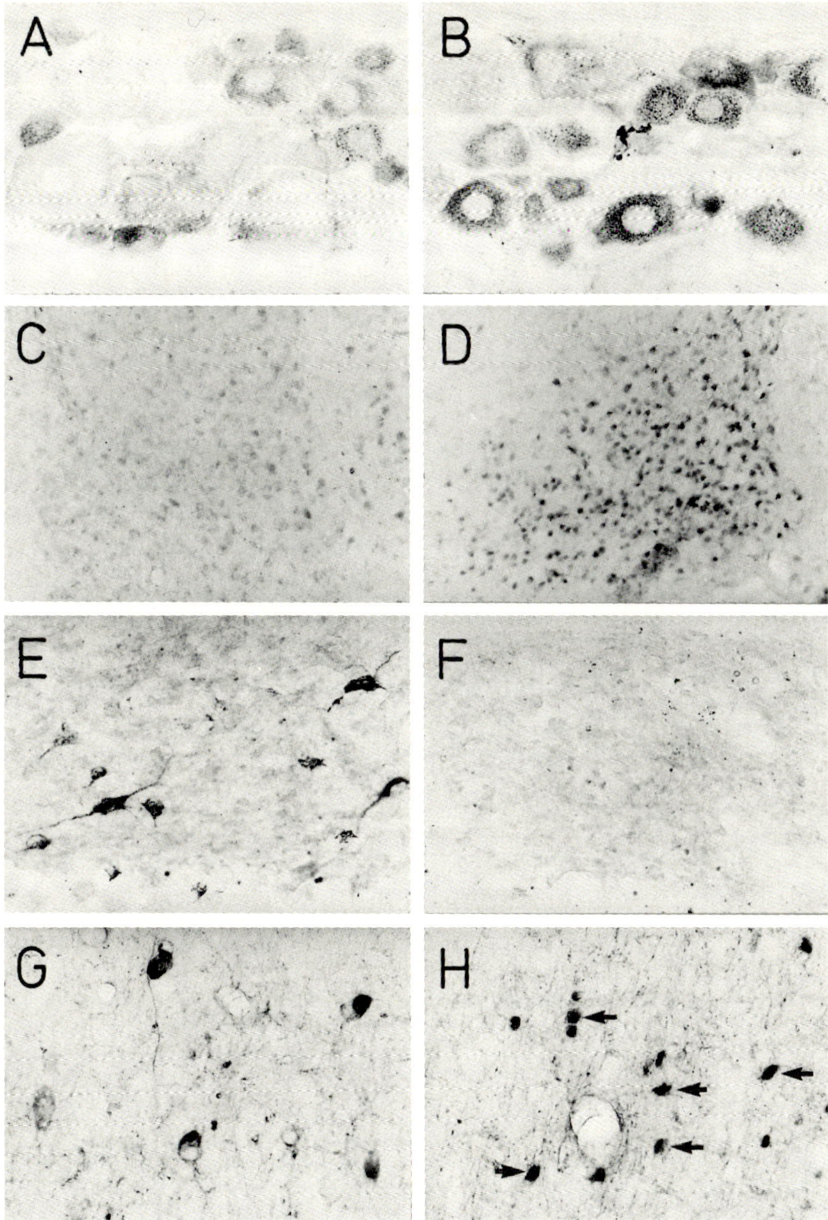

Fig. 5. Changes of nitric oxide synthase immunoreactivity (NOS-IR) in axotomized neurons in the adult rat. NOS-IR (A) in intact neurons of L4 dorsal root ganglia and (B) its *increase* 10 days following sciatic nerve cut. NOS-IR (C) in intact neurons of mammillary body and (D) its *increase* 30 days following transection of mammillo-thalamic tract. NOS-IR (E) in intact neurons of the cerebellar granule cells and (F) its *decrease* 5 days following transection of cerebellar granular cell-layer. NOS-IR (G) in intact neurons of medial septal ncl. and (H) its *decrease* 10 days following transection of fornix-fimbria. In (H), arrows mark c-Jun labelled neurons exhibiting only a residual cytoplasmic NOS-IR.

nerve increases the c-Jun-IR in primary sensory neurons and motoneurons (Herdegen et al. 1991b; Leah et al. 1991, 1993). Similarly, injection of colchicine or vinblastine into the thalamus induces c-Jun and Jun D, but not c-Fos and Fos B, expression in the affected area (Leah et al. 1993). These findings suggest that c-Jun might be induced by a negative signal, the omission of molecules that indicate the intactness of the neuron-target-axis under normal conditions.

Complete, not partial axotomy induces c-Jun expression.
Transection of the hippocampal commissure and the corpus callosum did not induce c-Jun expression in the damaged neurons of hippocampal and cortical neurons which have extented axon collaterals (Leah et al. 1993). These findings suggest that transection of axon collaterals is not effective for c-Jun expression.

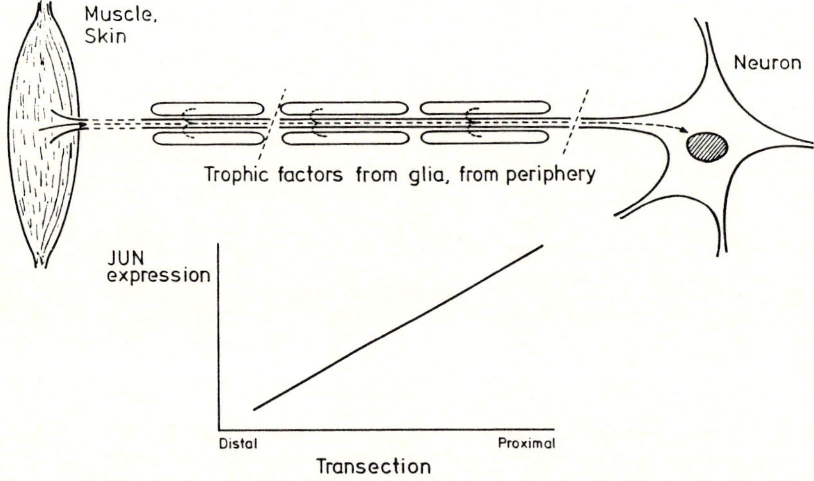

Fig. 6. Hypothesis for the induction of c-Jun following peripheral nerve section by a "target-derived suppressor": the shorter the transected proximal nerve stump, the lower the amount of putative trophic factors released by glial cells which could supply trophic support for the injured neuron. The proximity of nerve transection to the axotomized perikaryon results in an increased c-Jun expression (see also Fig. 1I-K).

Does the phosphorylation of CREB trigger the different patterns of IEG expression following axotomy and transsynaptic stimulation?

CREB is involved in the regulation of c-Jun expression. Phosphorylated CREB (p-CREB) evokes transcription of *c-jun*, whereas dephosphorylated CREB acts as suppressor (Lamph et al. 1990). Therefore, it is of interest whether axotomy induces changes in the phosphorylation of CREB. Very recently, it was reported that phosphorylation of CREB induces c-Fos expression to a much higher extent compared to the expression of c-Jun (Peunova and Enikopolov 1993). It is our hypothesis that transsynaptic stimulation induces phosphorylation of CREB, which triggers the subsequent activation of transcriptional operations dominated by c-Fos and Jun B; in contrast, axotomy induces c-Jun in a CREB-independent manner, and, moreover, blocks the constitutive expression of CREB. In collaboration with D. Ginty and M. Greenberg, we will analyze the phosphorylation of CREB following axotomy and transsynaptic stimulation (Ginty et al. 1993).

Expression of c-Jun is not involved in physiological growth

The induction of Jun proteins in axotomized neurons and their repression following axonal elongation with successful re-estabishment of the neuron-target-axis also points out a role of Jun in physiological growth of neurons. However, the patterns of Jun expression in growing neurons do not differ from that of resting neurons. Thus, Jun-IR is absent in the ganglion cells of the margin of goldfish retina (Herdegen et al. 1993a) which undergo mitosis and axonal elongation of their fibers contributing to the optic nerve; the percentage of DRG neurons with c-Jun-IR in embyronic rats (embryonic day E14, E18 and E20) does not differ from that of adult rats (Fig. 7A) (Utzschneider, Herdegen, Koscic and Waxman, unpublished observations). Similar, expression of c-*jun* mRNA is restricted to a low number of neuronal populations in the mouse nervous system during development (Wilkinson et al. 1989).

Expression of Jun, Fos and Krox proteins in neurons of adult dorsal root ganglia *in vitro*

In contrast to the axotomy of DRG neurons *in vivo*, explantation of adult rat DRG and their cultivation in polyornithin-covered culture dishes induces within 24 h the expression of c-Jun, Jun B, Jun D, c-Fos, Fos B, Krox-20 and Krox-24 in virtually all explanted neurons (Fig. 7B-D) (in collaboration with J. Koscic and S. Waxman; Herdegen et al. 1993f). Cell culture experiments also alter the intraneuronal localization of transcription factor proteins: Krox-20 shows a distinct cytoplasmic IR in DRG of untreated rats that is not altered by axotomy (Herdegen et al. 1993d); *in vitro*, however, Krox-24 is translocated in the nucleus of explanted neurons (Fig. 7E,F). Between 4-7 days, the expression of IEGs decline more rapidly in large diameter neurons compared to small diameter neurons. c-Jun and Jun D show a more sustained up-regulation compared to c-Fos and Jun B. This indicates that following an equilibrium and adjustment of metabolism of explanted neurons, the expression patterns of IEGs re-approach to those present *in vivo* following axotomy.

Expression of Jun, Fos and Krox proteins in glial cells of transected nerve fibers

Induction of IEG does not occur in reactive microglia around axotomized neurons in contrast to CREB protein (Herdegen et al. 1991b). However, specific patterns

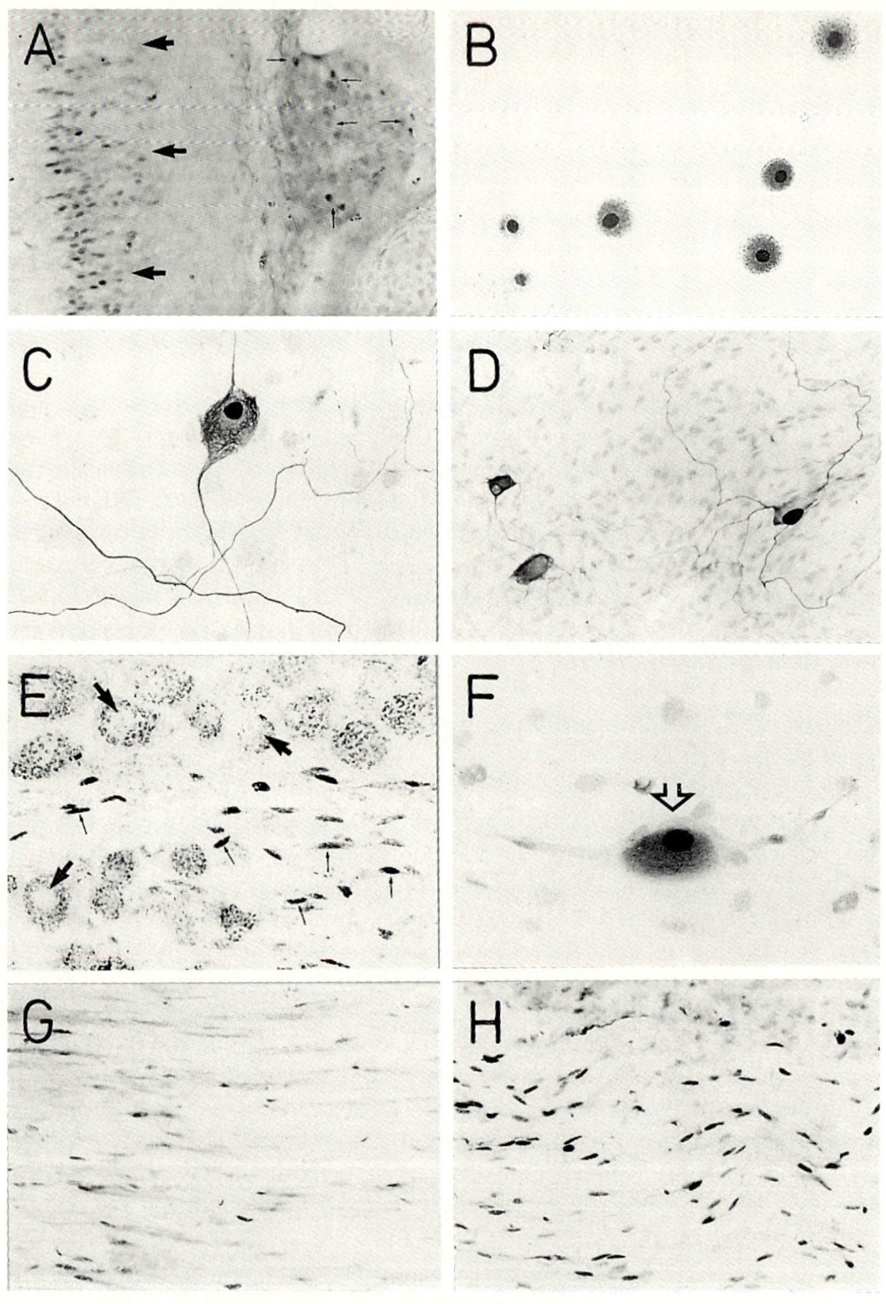

of IEG up-regulation can be observed in the distal nerve stump following transection of peripheral nerve fibers.

Within 3 h and 12 h, the number of glial cells labelled by c-Jun, Jun D, c-Fos and Krox-24 increases in the distal but not proximal nerve stump (Fig. 7G,H) (Voss and Herdegen, unpublished observations). These glial IR showed a different persistence lasting up to 20 days. In contrast, transection of central nerve fiber tract such as the corpus callosum, induced a transient expression of IEG in glial cells up to 7 days which was restricted to the sites adjacent to the transection. Interestingly, Jun B appeared in glial cells of central nerve fibers but was almost absent in glial cells of peripheral nerve stump.

In vitro experiments have attributed a role to c-Jun and c-Fos expression in the regulation of the NGF gene (Hengerer et al. 1990). However, the transient expression of c-Fos and the more lasting up-regulation of NGF expression by glial cells suggest that following the decline of c-Fos, the control of NGF expression has to be exerted by transcription factors different from the classical AP-1 complex.

Differential expression of Jun, Fos and Krox proteins in axotomized, deafferented and mechanically injured neurons

For the evaluation of the functional role of IEG encoded proteins it is important to define those stimuli which induce IEGs and their protein products. Deafferentation, transsynaptic impulse discharge and axotomy can occur in one neuroanatomical compartment. Thus, c-Fos and Jun B are expressed in spinal cord neurons following sciatic nerve cut, and in the substantia nigra reticularis following MFB transection. It could be demonstrated that the expression of c-Fos and Jun B was due to transsynaptic stimulation and not due to axotomy (Herdegen et al. 1993b; Leah et al. 1991; Weiser et al. 1993).

Mechanical lesions of the cortex, hippocampus and thalamus by stereotaxic introduction of a blade evoke different patterns of IEG expression compared to axotomy: all Jun, Fos and Krox-24 proteins are induced in neurons within 3 h and differentially persisted up to 5 days in the area surrounding the site of lesion. This might be due to spontaneous impulse discharge by the introduction of the blade, and due to transmembranous stimulation by inflammatory and humoral agents from the wound rather than due to axotomy.

← **Fig. 7.** (A) c-Jun in neurons of dorsal root ganglia (small arrows) and preganglionic sympathetic neurons of IML dorsal horn (big arrows) in an embryonic rat (E18). Transversal cut of spinal cord (Utzschneider, Herdegen, Koscic and Waxman, unpublished observations). (B) c-Jun, (C) Jun B and (D) c-Fos in adult dorsal root ganglia (DRG) (B) 24 h, (C,D) 7 days after explantation and cell culture cultivation. In (C,D), neurons were counterstained against the neurofilament protein SMI that exclusively labels neurons. (E) Axotomy does not change the cytoplasmic immunoreactivity of Krox-20 in axotomized neurons of lumbar DRG neurons (big arrows) and the nuclear labelling of glial cells (small arrows) following sciatic nerve cut whereas (F) following explantation of lumbar DRG neurons into cell culture system, Krox-20 is translocated into the nucleus within 24 h (arrow) (for B-D,F see Herdegen et al. 1993f). c-Jun-IR in glial cells in the (G) proximal and (H) distal nerve stump 3 days following sciatic nerve cut.

Discussion

The lasting and selective expression of c-Jun and Jun D following peripheral nerve cut, and of c-Jun, Jun D and Krox-24 following transection of central nerve fibers, represents a novel pattern of IEG expression in the nervous system. The principle finding of c-Jun and Jun D expression apparantly represents a general neuronal reaction due to axotomy. This expression pattern is unique in the view that numerous pathophysiological stimuli evoke a rapid and fairly transient cascade of IEG expression comprising the parallel *de novo* synthesis of c-Jun, Jun B, Jun D, c-Fos, Fos B, Krox-24, and, following particular experimental paradigms, also of Krox-20 (Herdegen et al. 1993d). These paradigms include noxious peripheral stimulation (Bullit 1990; Herdegen et al. 1991a,c,d, 1994; Hunt et al. 1987; Lanteri-Minet et al. 1993; Pertovaara et al. 1993; Tölle et al. 1994; Williams et al. 1990; Wisden et al. 1990), epileptic seizures (Gass et al. 1992a, 1993b; Cole et al. 1989; Morgan et al. 1987; Wisden et al. 1990), cerebral ischemia (Gass et al. 1992b; Kiessling et al. 1993; Nowak et al. 1993), cortical spreading depression (Herdegen et al. 1993e), shift of circadian rhythms (Rusak et al. 1990), memory formation (Anoukhin and Rose 1991; Heurteaux et al. 1993; Rose 1991) as well as pharmacological treatments (Bing et al. 1991; Gass et al. 1993a; Lebrun et al. 1994; Pertovaara et al. 1993; Tölle et al. 1994). Similarly, numerous *in vitro* experiments have demonstrated the parallel co-expression of Jun, Fos and Krox (Almendral et al. 1988; Bravo 1990; Kovary and Bravo 1991a,b, 1992). Following transsynaptic input, selectivity, if at all, was found for induction of c-*fos without* c-*jun* following membrane depolarization of fibroblasts (Bartel et al. 1989) or following cortical spreading depression (Herdegen et al. 1993f).

The enhanced levels of c-Jun proteins are due to persisting up-regulation of c-*jun* gene as demonstrated by *in situ* hybridization and blotting (Haas et al. 1993; Jenkins et al. 1991, 1993b; Koistinaho et al. 1993a-b; Rutherfurd et al. 1992), whereas the levels of c-*fos* mRNA are not affected by axotomy (Haas et al. 1993; Weiser et al. 1993). In contrast, transiently elevated levels of *jun* B mRNA (Haas et al. 1993) are not translated into the Jun B protein (Haas and Herdegen, unpublished observation). This block of translation might contribute to the absence of Jun B protein in axotomized neurons investigated so far in our laboratory (Herdegen et al. 1991b, 1992a, 1993a,b). The block of *jun* B translation has important consequences for the transcriptional operations: Jun B protein acts as a functional antagonist of c-Jun protein (Deng and Karin 1993; Schütte et al. 1989), and the absence of Jun B in axotomized neurons might underly the supposed regulatory dominance of c-Jun in the neuronal cell body-response. Recent observations by Schlingensiepen et al. (1992; see also this volume) have demonstrated that Jun B and c-Jun differ in their biological meaning for neuronal differentiation and sprouting.

Dissociation of mRNA and protein expression was also found in our laboratory following application of UV-irradiation on one hindpaw of rats; this experimental procedure increased the contents of *jun* B and *jun* D mRNA in the spinal dorsal horn (measured by PCR) but this increase was not followed by increase in Jun Band Jun D-IR (Gillardon et al. 1992).

In addition to the known neuroanatomical localization of axotomized neurons, there is also experimental evidence that c-Jun is virtually expressed in axotomized neurons, e.g. by double-labelling of c-Jun labelled neurons with the retrograde tracers fast blue (Leah et al. 1991) and HRP-coupled gold (Brecht et al. 1994).

The mechanisms for the selective expression of c-Jun without c-Fos and Jun B are still enigmatic. The delayed onset of Jun expression between 5 and 24 hours (Haas et al. 1993; Herdegen et al. 1991b, 1992a, 1993a,b; Jenkins et al. 1992, 1993; Koistinaho et al. 1993a,b) after nerve transection apparently excludes signal transfer by electrical impulse propagation, but suggests an axonal signal transfer that is dependent on the distance between the site of lesion and the nuclear Jun expression.

The intracellular and intranuclear cascade of second and third messengers contributing to the *de novo* synthesis of c-Jun but not of c-Fos remains also to be elucidated. However, findings about the expression and phosphorylation of the CREB transcription factor suggest that axotomy and transsynaptic neuronal stimulation have different effects on the function and expression of this constitutively expressed transcription factor: following nerve cut, CREB-IR is decreased up to complete disappearance in axotomized neurons; moreover, axotomy is not effective in changing the immunoreactivity of phosphorylated CREB (Herdegen, Ginty and Greenberg, unpublished observation). On the other hand, transsynaptic stimulation evokes an early phosphorylation of CREB (Ginty et al. 1993) and does not change the level of CREB expression (Herdegen et al. 1993e). Very recently, it has been demonstrated that phosphorylation of CREB is much more effective for induction of *c-fos* compared to induction of *c-jun* (Peunova and Enikolopov 1993).

Changes in expression of transcription factors are confined to a particular neuronal reaction

Our comprehensive observations suggest that the selective expression of c-Jun and Jun D seems to be related to a particular neuronal reaction: to axotomy with a subsequent substantial increase of *de novo* protein synthesis. Nerve fiber lesions with weak or absent cell body-reaction and minor changes of protein synthesis do not induce Jun expression in the affected perikaryon, e.g. following transection corpus callosum (Leah et al. 1993) and dorsal rhizotomy (Jenkins et al. 1993b). In addition, collateral sprouting can also result in Jun expression. Thus, sciatic nerve cut increases the c-Jun-IR in the contralateral intact dorsal root ganglia (Wickert et al. unpublished observation) and in the adjacent primary afferent neurons of saphenous nerve (Jenkins et al. 1993b). Moreover, systemic application of NMDA-recpetor antagonist MK-801 stimulating the sprouting of primary afferent fibers, is effective in c-Jun induction in dorsal root ganglia, e.g. of sciatic nerve (Leah et al. unpublished observation). On the other hand, physiological growth is apparently not effective for c-Jun up-regulation (Herdegen et al. 1993a; Wilkinson et al. 1988; Utzschneider and Herdegen, unpublished observation). These findings suggest that the selective c-Jun activation without concomitant c-Fos expression represents a specific pathophysiological reaction in neurons due to nerve fiber transection and to compensatory sprouting. These particular transcriptional operations might override the physiological cellular programs of developing and growing neurons as well as of intact adult neurons. The selective activation of Jun proteins underlying a particular neuronal reaction might offer an explanation to the observation that the alterations of protein synthesis evoked by axotomy only to some extent mimick those present during development and growth (Miller et al. 1989; Strittmatter et al. 1992; Tetzlaff et al. 1991).

For the evaluation of those neuronal processes controlled by *de novo* expression of IEG encoded transcription factors, it has to be considered that IEGs might not be involved in neuronal processes predominantly triggered by local events within synaptic terminals and growth cones.

Supported by Deutsche Forschungsgemeinschaft, grant Zi 110/22-1.

References

Almendral JM, Sommer D, MacDonald-Bravo H, Burckhardt J, Perera J, Bravo R (1988) Complexity of the early genetic response to growth factors in mouse fibroblasts. Mol Cell Biol 8:2140-2148

Angel P, Allegretto EA, Okino ST, Hattori K, Boyle WJ, Hunter T, Karin M (1988) Oncogene *jun* encodes a sequence-specific trans-activator similar to AP-1. Nature 332:166-171

Angel P, Karin M (1991) The role of Jun, Fos and the AP-1 complex in cell-proliferation and transformation. Biochim Biophys Acta 1072:129-157

Anoukhin KV, Rose SPR (1991) Learning-induced increase of immediate-early gene messenger RNA in the chick forebrain. Eur J Neurosci 3:162-167

Bartel DP, Sheng M, Lau LF, Greenberg ME (1989) Growth factors and membrane depolarization activate distinct programs of early response gene expression: dissociation of *fos* and *jun* induction. Gene Develop 3:304-313

Bing G, Filer D, Miller JC, Stone EA (1991) Noradrenergic activation of immediate-early genes in rat cerebral cortex. Mol Brain Res 11:43-46

Björklund A, Katzman R, Stenevi U, West KA (1971) Development and growth of axonal sprouts from noradrenaline and 5-hydroxytryptamine neurons in the rat spinal cord. Brain Res 31:21-33

Bravo R (1990) Growth factor inducible genes in fibroblasts. In: Habenicht A (ed) Growth factors, Differentiation Factors and Cytokines. Springer-Verlag, Berlin-Heidelberg, pp 324-343

Brecht S, Gass P, Anton F, Bravo R, Zimmermann M, Herdegen T (1994) Induction of c-Jun and suppression of CREB transcription factor proteins in axotomized neurons of substantia nigra, and covariation with tyrosine hydroxylase. Mol Cell Neurosc, in press

Bredt DS, Snyder SH (1992) Nitric oxide, a novel neuronal messenger. Neuron 8:3-11

Bullitt E (1990) Expression of c-Fos like protein as a marker for neuronal activity following noxious stimulation in the rat. J Comp Neurol 296:517-530

Cole AJ, Saffen DW, Baraban JM, Worley PF (1989) Rapid increase of an immediate-early gene messenger RNA in hippocampal neurons by synaptic NMDA receptor activation. Nature 340:474-476

Deng T, Karin M (1993) Jun B differs from c-Jun in its DNA-binding and dimerization domains, and represses c-Jun by formation of inactive heterodimers. Genes Development 7:479-490

Fawcett JW (1992) Intrinsic neuronal determinants of regeneration. TINS 15:5-8

Fiallos-Estrada CE, Herdegen T, Kummer W, Mayer W, Bravo R, Zimmermann M (1993) Long-lasting increase of nitric oxide synthase immunoreactivity and NADPH-diaphorase reaction, and co-expression with the nuclear c-Jun protein in rat dorsal ganglion neurons following sciatic nerve transection. Neurosci Lett 150:169-173

Formann DS, McQuarrie IG, Labore FW, Wood DJ, Stone LS, Braddock CH, Fuchs DA (1980) Time course of the conditioning lesion effect on axonal regeneration. Brain Res 182:180-185

Gass P, Herdegen T, Bravo R, Kiessling M (1992a) Induction of immediate-early gene encoded proteins in the rat hippocampus after bicuculline-induced seizures: differential expression of Krox-24, Fos and Jun proteins. Neurosci 48:315-324

Gass P, Spranger M, Herdegen T, Köck P, Bravo R, Hacke W, Kiessling M (1992b) Induction of Fos and Jun proteins after focal ischemia in the rat: differential effect of the N-methyl-D-aspartate receptor antagonist MK-801. Acta Neuropathol 84:545-553

Gass P, Herdegen T, Bravo R, Kiessling M (1993a) Induction and suppression of immediate-early genes (IEGs) in specific rat brain regions by the non-competitive NMDA receptor antagonist MK-801. Neurosci 53:749-758

Gass P, Herdegen T, Bravo R, Kiessling M (1993b) Induction of six immediate-early gene encoded proteins in the rat brain after kainic acid induced limbic seizures: effects of NMDA receptor antagonist MK-801. Eur J Neurosci 5:933-943

Gillardon F, Herdegen T, Bravo R, Zimmermann M (1992) Regulation of c-*fos* and *jun* B mRNA translation in rat spinal neurons following cutaneous formalin injection or UV-B irradiation: information storage at the mRNA level? Abstr Soc Neurosci 18:135

Ginty DD, Kornhauser JM, Thompson MA, Bading H, Mayo KE, Takahashi JS, Greenberg ME (1993) Regulation of CREB phosphorylation in suprachiasmatic nucleus by light and a circadian clock. Science 260:238-241

Grafstein B (1986) The retina as a regenerating organ. In: Adler R, Farber D (eds) The Retina: a Model for Cell Biology Studies, Academic Press, New York, pp 275-335

Haas CA, Donath C, Kreutzberg GW (1993) Differential expression of immediate-early genes after transection of the facial nerve. Neurosci 53:91-99

Hengerer B, Lindholm D, Heumann R, Rüther U, Wagner EF, Thoenen H (1990) Lesion-induced increase in nerve growth factor mRNA is mediated by c-fos. Proc Natl Acad Sci USA 87:3899-3903

Herdegen T, Leah J, Zimmermann M (1989) Expression of the Jun proto-oncogene encoded protein in spinal neurons. Eur J Neurosci, Suppl 2:104

Herdegen T, Leah J, Bravo R (1990) Different expression of Jun, Fos and Krox-24 proteins in adult CNS: dependence on neurobiological events. J Cell Biochemistry, Suppl 14F:310

Herdegen T, Kovary K, Leah JD, Bravo R (1991a) Specific temporal and spatial distribution of Jun, Fos and Krox-24 proteins in spinal neurons following noxious transynaptic stimulation. J Comp Neurol 313:178-191

Herdegen T, Kummer W, Fiallos-Estrada CE, Leah JD, Bravo R (1991b) Expression of c-Jun, Jun B and Jun D in the rat nervous system following transection of the vagus nerve and cervical sympathetic trunk. Neurosci 45:413-422

Herdegen T, Leah JD, Manisali A, Bravo R, Zimmermann M (1991c) c-Jun-like immunoreactivity in the CNS of the adult rat: basal and transynaptically induced expression of an immediate-early gene. Neurosci 41:643-654

Herdegen T, Tölle T, Bravo R, Zieglgänsberger W, Zimmermann M (1991d) Sequential expression of Jun B, Jun D and Fos B proteins in rat spinal neurons: cascade of transcriptional operations during nociception. Neurosci Lett 129:221-224

Herdegen T, Fiallos-Estrada CE, Schmid W, Bravo R, Zimmermann M (1992a) The transcription factors c-Jun, Jun D and CREB, but not Fos and Krox-24, are differentially regulated in axotomized neurons following transection of rat sciatic nerve. Mol Brain Res 14:155-165

Herdegen T, Fiallos-Estrada CE, Schmid W, Bravo R, Zimmermann M (1992b) The transcription factor CREB, but not IEG encoded proteins, is expressed in activated microglia of lumbar spinal cord following sciatic nerve transection in the rat. Neurosci Lett 142:57-61

Herdegen T, Bastmeyer M, Bähr M, Bravo R, Stürmer CAO, Zimmermann M (1993a) Expression of Jun, Krox and CREB transcription factors in goldfish and rat ganglion cells following optic nerve lesions is related to axonal sprouting. J Neurobiol 24:528-543

Herdegen T, Brecht S, Mayer W, Leah JD, Kummer W, Bravo R, Zimmermann M (1993b) Long-lasting expression of Jun and Krox transcription factors and nitric oxide synthase in intrinsic neurons of the rat brain following axotomy. J Neurosci 13:4130-4146

Herdegen T, Fiallos-Estrada CE, Bravo R, Zimmermann M (1993c) Colocalisation and covariation of the nuclear c-Jun protein with galanin in primary afferent neurons and with CGRP in spinal motoneurons following transection of rat sciatic nerve. Mol Brain Res 17:147-154

Herdegen T, Kiessling M, Bravo R, Zimmermann M, Gass P (1993d) The Krox-20 transcription factor in the adult brain: novel expression pattern of an immediate-early gene encoded protein. Neurosci 57:41-51

Herdegen T, Sandkühler J, Gass P, Kiessling M, Bravo R, Zimmermann M (1993e) Jun, Fos, Krox and CREB transcription factor proteins in the rat cortex: basal expression and expression by cortical spreading depression and epileptic seizures. J Comp Neurol 333:271-288

Herdegen T, Bravo R, Waxman S, Kocsis J (1993f) Jun, Fos, Krox and CREB protein expression is different in adult dorsal root ganglion neurons studied *in vivo* and *in vitro*. Abstr Soc Neurosci 19:1729

Herdegen T, Rüdiger S, Mayer B, Bravo R, Zimmermann M (1994) Expression of nitric oxide synthase and colocalization of Jun, Fos and Krox proteins in spinal neurons following noxious peripheral stimulation. Mol Brain Res, in press

Herschman HR (1991) Primary response genes induced by growth factors and tumor promoters. Annu Rev Biochem 60:281-319

Heurteaux C, Messier C, Destrade C, Lazdunski M (1993) Memory processing and apamin induce immediate-early gene expression in mouse brain. Mol Brain Res 3:17-22

Hunt SP, Pini A, Evan G (1987) Induction of c-Fos-like protein in spinal cord neurons following sensory stimulation. Nature 328:632-634

Jenkins R, Hunt SP (1991) Long-term increase in the levels of c-*jun* mRNA and Jun protein-like immunoreactivity in motor and sensory neurons following axon damage. Neurosci Lett 129:107-110

Jenkins R, O'Shea R, Thomas KL, Hunt SP (1993a) c-Jun expression in substantia nigra neurons following striatal 6-hydroxydopamine lesions in the rat. Neurosci 53:447-455

Jenkins R, Tetzlaff W, Hunt SP (1993b) Differential expression of immediate-early genes in rubrospinal neurons following axotomy in rat. Europ J Neurosci 5:203-209

Kiessling M, Stumm G, Xie Y, Herdegen T, Aguzzi A, Bravo R, Gass P (1993) Differential transcription and translation of immediate early genes in the gerbil hippocampus after transient global ischemia. J Cereb Blood Flow Metab, in press

Koistinaho J, Hicks KJ, Sagar SM (1993a) Long-term induction of c-*jun* mRNA and Jun protein in rabbit retinal ganglion cells following axotomy or colchicine treatment. J Neurosci Res 34:250-255

Koistinaho J, Pelto-Huikko, Sagar SM, Dagerlind A, Roivainen R, Hökfelt T (1993b) Injury-induced long-term expression of immediate-early genes in the rat superior cervical ganglion. NeuroReport 4:37-40

Kovary K, Bravo R (1991a) The Jun and Fos protein families are both required for cell cycle progression in fibroblasts. Mol Cell Biol 11:4466-4472

Kovary K, Bravo R (1991b) Expression of different Jun and Fos proteins during the G0 to G1 transition in mouse fibroblasts: *in vitro* and *in vivo* associations. Mol Cell Biol 11:2451-2459

Kovary K, Bravo R (1992) Existence of different Fos/Jun complexes during the G0-to-G1 transition and during exponential growth in mouse fibroblasts: differential role of Fos proteins. Mol Cell Biol 12:5015-5023

Lamph WW, Dwarki VJ, Ofir R, Montminy M, Verma IM (1990) Negative and positive regulation by transcription factor cAMP response element-binding protein is modulated by phosphorylation. Proc Natl Acad Sci USA 87:4320-4324

Lanteri-Minet M, de Pommery J, Herdegen T, Weil-Fugazza J, Bravo R, Menetrey D (1993) Differential time-course and spatial expression of Fos, Jun and Krox-24 proteins in spinal cord of rats undergoing subacute or chronic somatic inflammation. J Comp Neurobiol 333:223-235

Leah JD, Herdegen T, Kovary K, Bravo R (1991) Selective expression of Jun proteins following peripheral axotomy and axonal transport block in the rat: evidence for a role in the regeneration process. Brain Res 566:198-207

Leah JD, Herdegen T, Dragunow M, Bravo R (1993) Differential expression of immediate-early gene proteins following axotomy and axonal transport block in the rat CNS. Neuroscience 57:53-66

Lebrun C, Blume A, Herdegen T, Bravo R, Unger T (1994) Angiotensin II induces differential actions of transcription factor complexes in the rat: expression of Jun, Fos and Krox proteins. Neuroscience, submitted

Lehmann S, Kuchler S, Gobaille S, Marschal P, Badache A, Vincendon G, Zanetta JP (1993) Lesion-induced re-expression of neonatal recognition molecules in adult rat cerebellum. Brain Res Bulletin 30:515-521

McKerracher L, Hirscheimer A (1992) Slow transport of the cytoskeleton after axonal injury. J Neurobiol 23:568-578

Miller FD, Tetzlaff W, Bisby MA, Fawcett JW, Milner RJ (1989) Rapid induction of the major embryonic alpha-tubulin mRNA, T alpha-1, during nerve regeneration in adult rats. J Neurosci 9:1452-1463

Moncada S, Palmer RMJ, Higgs EA (1991) Nitric oxide: physiology, pathophysiology and pharmacology. Pharmacol Rev 43:109-142

Montminy MR, Gonzalez GA, Yamamoto K (1990) Regulation of cAMP-inducible genes by CREB. TINS 13:184-188

Morgan JI, Cohen DR, Hempstead JL, Curran T (1987) Mapping patterns of c-Fos expression in the central nervous system after seizure. Science 237:192-196

Morgan JI, Curran T (1991) Stimulus-transcription coupling in the nervous system: involvement of the inducible proto-oncogenes *fos* and *jun*. Annu Rev Neurosci 14:421-451

Neiss WF, Guntinas-Lichius O, Angelov DN, Gunkel A, Stennert E (1992) The hypoglossal-facial anastomosis as model of neuronal plasticity in the rat. Ann Anat 174:419-433

New IIV, Mudge AW (1986) Calcitonin gene-related peptide regulates muscle acetylcholine receptor synthesis. Nature 323:809-811

Nowak Jr. TS, Osborne OC, Suga S (1993) Stress protein and proto-oncogene expression as indicators of neuronal pathophysiology after ischemia. Progress in Brain Res 96:195-208

Pertovaara A, Bravo R, Herdegen T (1993) Induction and suppression of immediate-early genes by selective alpha-2-adrenoceptor agonist and antagonist in the rat brain following noxious peripheral stimulation. Neurosci 54:117-126

Peunova N, Enikolopov G (1993) Amplification of calcium-induced gene transcription by nitric oxide in neuronal cells. Nature 364:450-453

Rose SPR (1991) How chicks make memories: the cellular cascade from c-*fos* to dendritic remodelling. TINS 14:390-397

Rusak B, Robertson HA, Wisden W, Hunt SP (1990) Light pulses that shift rhythms induce gene expression in the suprachiasmatic nucleus. Science 248:1237-1240

Rutherfurd SD, Louis WJ, Gundlach AL (1992) Induction of c-Jun expression in vagal motoneurons following axotomy. NeuroReport 3:465-468

Ryder K, Lau LF, Nathans D (1988) A gene activated by growth factors is related to the oncogene v-jun. Proc Natl Acad Sci USA 85:1487-1491

Ryseck R, Hirai S, Yaniv M, Bravo R (1988) Transcriptional activation of c-JUN during the G0/G1 transition in mouse fibroblasts. Nature 334:535-537

Sagar SM, Sharp FR, Curran T (1988) Expression of c-Fos protein in the brain: metabolic mapping at the cellular level. Science 240:1328-1331

Schlingensiepen KH, Gerdes W, Seifert W, Brysch W (1992) Role of Jun B in neuronal differentiations. Eur J Neurosci Suppl 5:162

Schütte J, Viallet J, Nau M, Segal S, Fedorko J, Minna J (1989) Jun B inhibits and c-Fos stimulates the transforming and trans-activating activities of c-Jun. Cell 59:987-997

Sheng M, McFadden G, Greenberg ME (1990) Membrane depolarization and calcium induce c-*fos* transcription via phosphorylation of transcription factor CREB. Neuron 4:571-582

Sheng M, Greenberg ME (1990) The regulation and function of c-*fos* and other immediate-early genes in the nervous system. Neuron 4:477-485

Skene JHP (1989) Axonal growth-associated proteins. Annu Rev Neurosci 12:127-156

Strittmatter SM, Vartanian T, Fishman MC (1992) GAP-43 as a plasticity protein in neuronal form and repair. J Neurobiol 23:507-520

Stürmer CAO, Bastmeyer M, Bähr M, Strobel G, Paschke K (1992) Trying to understand axonal regeneration in the CNS of fish. J Neurobiol 23:537-550

Tetzlaff W, Alexander SW, Miller FD, Bisby MA (1991) Response of facial and rubrospinal neurons to axotomy: changes in mRNA expression for cytoskeletal proteins and GAP-43. J Neurosci 11:2528-2544

Tölle TR, Herdegen T, Schadrack J, Bravo R, Zimmermann M, Zieglgänsberger W (1994) Application of morphine prior to noxious stimulation differentially modulates expression of Fos, Jun and Krox-24 in rat spinal cord neurons. Neurosci 58:305-321

Verge VKM, Xu Z, Xu XJ, Wiesenfeld-Hallin S (1992) Marked increase in nitric oxide synthase mRNA in rat dorsal root ganglia after peripheral axotomy: *in situ* hybridization and functional studies. Proc Natl Acad Sci USA 89:11617-11621

Villegas-Perez MP, Vidal-Sanz M, Raminsky M, Bray GM, Aguayo AJ (1992) Rapid and protracted phases of retinal ganglion cell loss follow axotomy in the optic nerve of adult rats. J Neurobiol 24:23-36

Vogt PK, Bos TJ (1990) Jun: oncogene and transcription factor. Adv Cancer Res 55:2-36

Weiser M, Baker H, Wessel TC, Joh TH (1993) Axotomy-induced differential gene induction in neurons of the locus ceruleus and substantia nigra. Mol Brain Res 17:319-327

Wilkinson DG, Bhatt S, Chavrier P, Bravo R, Charnay P (1989) Segment-specific expression of a zinc finger gene in the developing nervous system of the mouse. Nature 337:461-464

Williams S, Evans GI, Hunt SP (1990) Changing patterns of c-Fos induction in spinal neurons following thermal cutaneous stimulation in the rat. Neurosci 36:73-81

Wisden W, Errington ML, Williams S, Dunnett SB, Waters C, Hitchcock D, Evan G, Bliss TV, Hunt SP (1990) Differential expression of immediate-early genes in the hippocampus and spinal cord. Neuron 4:603-614

Wu W (1993) Expression of nitric oxide synthase (NOS) in injured CNS neurons as shown by NADPH diaphorase histochemistry. Exper Neurol 120:153-159

Xu XJ, Wiesenfeld-Hallin Z, Villar MJ, Fahrenkrug J, Hökfelt T (1990) On the role of galanin, substance P and other neuropeptides in primary sensory neurons in the rat: studies on spinal reflex excitability and peripheral axotomy. Eur J Neurosci 2:733-743

Immediate-early genes and opioid peptides

J. Kraus, B. Bacher, X. Wang, and V. Höllt
Institute of Physiology, University of Munich, Pettenkoferstraße 12, D-80336
München, Germany

Introduction

Cellular stimulation causes the rapid appearance of proteins in the nucleus which
function as signal-regulated transcription factors converting membrane events into
long-term changes in gene expression. These transcription factors are the products
of a variety of immediate early genes (IEG). One important group of these signal-
regulated transcription factors are the BZip proteins which contain each a basic
(B) and a leucine zipper (Zip) domain that are required for DNA binding and
dimerization, respectively.

The best studied members of this superfamily are the AP-1 (Jun/Fos) and
CREB/ATF proteins that control gene expression by binding to the TPA (12-O-
tetradecanoylphorbol-13-acetate)- response element (TRE) and cAMP-response
element (CRE), respectively (Karin 1989). The *jun* and *fos* gene families encode
four known Fos proteins (c-Fos, Fos B, Fra-1 and Fra-2) and three Jun proteins
(c-Jun, Jun B, Jun D). The Fos proteins form heterodimers with Jun-related proteins but do not form stable homodimers. The Jun proteins form homodimers and
heterodimers with each other, and heterodimers with Fos-related proteins (Morgan
and Curran 1991).

The CREB/ATF gene family also encodes a large variety of transcription factors, which also bind to DNA as dimers. Some of the members can even form
heterodimers with those of the Fos/Jun proteins which then bind preferentially to
CRE's (Hai and Curran 1991). CREB and other members of this gene family are
not immediate early genes in a strict sense, since they are, in general, not induced,
but bind constitutively to the CRE. Phosphorylation of CREB is thought to stimulate its transcriptional activity without altering its binding properties. In addition,
the immediate early response involves many other inducible transcription factors
as for example zinc-finger transcription factors like *egr*-1 (Sukhatme et al. 1988).

There is mounting evidence that transcription factors of the Jun/Fos or the
CREB families transactivate the three opioid peptide genes which encode proenkephalin, prodyndorphin and proopiomelanocortin. All three genes contain DNA
sequences that may serve as binding sites for these factors. In fact, there are many
physiological events, such as stress, pain, convulsions etc. which cause an increase in the expression of opioid peptide genes, which are, in general, preceded
by an increase in the expression of certain immediate early genes, such as c-*fos*
and/or c-*jun*.

The following short review summarizes and discusses data in literature pertinent
to this issue.

Proenkephalin

The proenkephalin gene encodes the precursor for the opioid peptides met- and leu-enkephalin. Proenkephalin is widely expressed in many tissues with highest concentrations in the adrenal medulla and in some brain areas, such as the striatum. Proenkephalin gene expression in the striatum is under negative control of dopamine. Thus, lesion of the dopaminergic nigro striatal pathway or the chronic blockade of dopamine receptors by D2-antagonists, such as halopcridol, results in a marked increase in striatal proenkephalin mRNA levels. In the hippocampus proenkephalin is expressed in low amounts in the granule cells of the dentate gyrus. Electrical or chemical activation of this area in rats results in long-term-potentiation or kindling and causes a marked induction of proenkephalin gene expression. Moreover, electrical stimulation of the trigeminus nerve elicits an induction of proenkephalin mRNA in neurons in lamina I and II in the nucleus caudalis of the trigeminal nuclear complex (Nishimori et al. 1989). All the stimuli which activate proenkephalin gene expression in the described areas of the central nervous system also enhance the expression of the immediate early genes c-*fos* and /or c-*jun* which precedes that of proenkephalin.

The second messenger systems involved in proenkephalin gene expression were studied in detail in cultured bovine adrenal medullary cells. It was shown that the activation of the three major second messenger pathways resulted in each case in an induction of proenkephalin gene expression in these cells. Thus, induction of proenkephalin gene expression can be accomplished by activation of either the cAMP/protein kinase A pathway, or the PIP_2-DAG/protein kinase C system, or the Ca^{2+}/membrane depolarization-dependent pathway. Activation of each of these signalling pathways also results in an early induction of mRNAs encoding c-Fos, c-Jun and Jun B which precedes the increase in proenkephalin mRNA levels (Stachowiak et al. 1990; Wang et al. 1993).

The sequence requirements for cAMP- and phorbol ester inducibility were initially tested by transfection of a human proenkephalin-CAT (chloramphenicol acetyltransferase)-reporter plasmid into monkey CV1 cells (Comb et al. 1986). A 37-bp DNA region located 107-71 bp 5' to the mRNA cap site confers both inducibility by cAMP and phorbol esters. Two DNA elements (ENKCRE-1 and ENKCRE-2) are located within this enhancer region: the proximal promoter element (ENKCRE-2) is essential for basal and regulated transcription. The distal promoter element (ENKCRE-1) has no inherent capacity to activate transcription, but synergistically augments cAMP and phorbol ester inducible transcription in the presence of ENKCRE-2. In addition, an AP-2 element was identified which is localized downstream of the ENKCRE-2 element. Inactivation of this element results in a five-fold reduction of both cAMP- and phorbol ester inducible transcription (Hyman et al. 1989). Stable transfection of the human proenkephalin-CAT reporter gene into C-6 glioma cells revealed that the ENKCRE-2 element also confers inducibility by calcium ions (Van Nguyen et al. 1990). Thus, ENKCRE-2 seems to be a key element in the regulation of proenkephalin gene expression. In fact, a minimal proenkephalin promoter consisting of multiple copies of ENKCRE-2 (minimum two) but not of multimers of ENKCRE-1 can confer cAMP, phorbol ester or calcium-dependent induction of transcription (Van Nguyen et al. 1990).

The ENKCRE-2 element contains the sequence TGCGTCA which differs by a single base from the consensus binding for AP-1 proteins TGAGTCA and by another single base from the consensus binding site for the cAMP-activated transcription factor CREB (TGACGTCA).

AP-1 transcription factors, such as heterodimers of c-Fos and c-Jun or Jun D proteins have been shown to directly bind the ENKCRE-2 element *in vitro* (Comb et al. 1988; Sonnenberg et al. 1989; Kobierski et al. 1991). In addition, Fos and c-Jun/Jun D heterodimers can transactivate proenkephalin gene expression when transfected in undifferentiated F9 cells (Sonnenberg et al. 1989; Kobierski et al. 1991).

On the other hand, cotransfection of CREB has also been shown to activate proenkephalin transcription in CV-1 cells (Huggenvik et al. 1991).

Since the *c-fos* gene itself contains a binding site for CREB (at position -60 of the human gene) it is possible that the activation of adenylate cyclase induces the proenkephalin gene either directly via CREB or indirectly via an induction of Fos containing AP-1 proteins (e.g. Fos/Jun heterodimers). The two possibilities may be distinguished by a) time-course differences and b) by protein synthesis inhibitors. If CREB directly binds to ENKCRE-2, the cAMP-mediated activation of proenkephalin gene expression should be fast and not be blockable by protein synthesis inhibitors. In fact, in C6 rat glioma cells which stably express a human proenkephalin-CAT fusion gene forskolin - an activator of the adenylate cyclase - induces a rapid induction (within one hour) of the proenkephalin-CAT fusion gene which cannot be inhibited by protein synthesis inhibitors such as anisomycin or cycloheximide (Kobierski et al. 1991). This indicates that CREB or another constitutive transcription factor binds directly to the proenkephalin gene. Induction of gene expression occurs by protein kinase A mediated phosphorylation of these factors. Such a constitutive factor which can be activated by protein kinase A may also be Jun D (Kobierski et al. 1991).

A completely different situation exists in bovine adrenomedullary chromaffin cells *in vitro*. In these cells forskolin or membrane depolarization by nicotine or high potassium causes a rapid induction of the expression of immediate early genes, such as *c-fos*, *c-jun* or *jun* B, whereas the induction of the proenkephalin gene occurs with a delay of three hours (Farin et al. 1990). Moreover, the induction of transcription of the proenkephalin gene and the increase in the level of proenkephalin mRNA by membrane depolarizing agents, forskolin or phorbol esters can be completely blocked by co-incubation of protein synthesis inhibitors, such as cycloheximide (Farin et al. 1990). This finding strongly suggests that the induction of the proenkephalin gene is indirectly mediated by AP-1 proteins which in turn may be activated by the phosphorylation of constitutive transcription factors such as CREB. This hypothesis is further strengthened by our recent data showing that cotransfection of a rat proenkephalin-CAT fusion gene (which contains a dimer of ENKCRE-2 in front of a minimal promoter) with a *c-fos* expression vector transactivates the expression of the proenkephalin-CAT fusion gene in primary cultured bovine chromaffin cells. In contrast, no transactivation of the proenkephalin-CAT fusion gene was obtained by cotransfection with a CREB expression vector. Moreover, coexpression of CREB with protein kinase A did not further increase the transcription of the proenkephalin-CAT fusion gene as did protein kinase A alone (Fig. 1). Finally, the induction of the proenkephalin gene expression in bovine chromaffin cells by forskolin and membrane depolarizing agents could be blocked by incubating the cells with c-*fos* antisense oligonucleotides (Wang, Bacher and Höllt, unpublished data).

As already mentioned, the proenkephalin gene can be induced in the central nervous system *in vivo* by a variety of manipulations. In the striatum the proenkephalin gene can be induced by destruction of the dopaminergic pathways or by application of D2-dopamine receptor antagonists, such as haloperidol. Interestingly, the application of haloperidol also results in an increase in the expression of c-*fos*. Moreover, an increase in the levels of proteins binding to the AP-1 site has been observed after haloperidol administration, indicating that the proenkephalin gene may be indirectly activated by AP-1 proteins. However, in a detailed band shift analysis Konradi and co-workers could clearly show that Fos and other proteins of the AP-1 complex do not bind to the CRE2-element of the proenkephalin promoter. Instead, they provided evidence that haloperidol caused a posphorylation of CREB in striatal cells *in vivo*. These findings strongly indicate that haloperidol activates proenkephalin gene expression not via expression ot the c-*fos* gene, but via phosphorylation of CREB (Konradi et al. 1993).

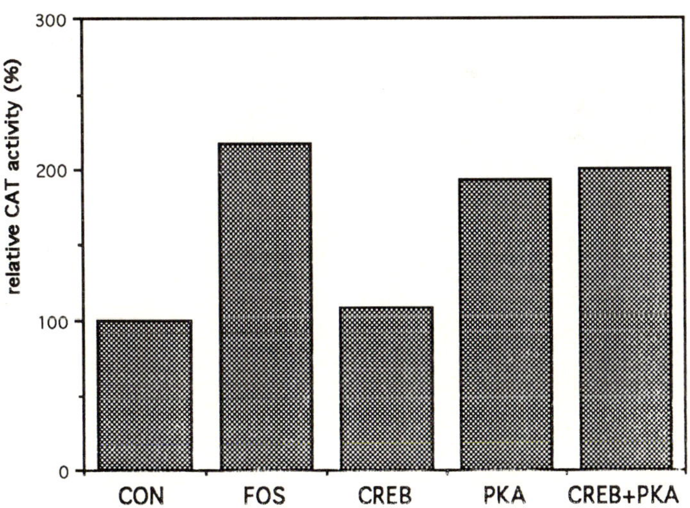

Fig. 1. Overexpression of c-Fos and protein kinase A, but not CREB, results in an increased transcriptional activity of PENK-CAT reporter genes in bovine chromaffin cells. Primary cultured adrenal medullary cells were transfected with PENKCAT(CRE-2)2 which contains two copies of ENKCRE-2 in front of sequences spanning from nt's -76 to +50 of the rat PENK gene (CON). Additionally, expression vectors for c-*fos* (FOS), CREB and protein kinase A (PKA) were co-transfected as indicated. The total amount of DNA was kept constant by addition of pRc/CMV.

Application of electrical and chemical stimuli increases immediate early genes and proenkephalin mRNA levels in the granule cells of the dentate gyrus of the hippocampus. Whether the increase in proenkephalin transcription is mediated via AP-1 proteins or via CREB is not clear. The fact that the time-course of proenkephalin gene expression is slower than that of c-*fos* and other immediate early genes appears to indicate that here AP-1 transcription factors may mediate proenkephalin gene expression.

In conclusion, these results indicate that the induction of proenkephalin gene expression varies in different cell types. In some cells it involves AP-1 transcription factors, in other cell types phosphorylated CREB appears to directly activate the proenkephalin gene.

Prodynorphin

The prodynorphin gene encodes a precursor protein which is cleaved into dynorphin and several related peptides which all have a high affinity for the k-opioid receptor. The gene is expressed in a variety of tissues, but more restricted than the proenkephalin gene. The regulation of the prodynorphin gene has been studied in the hypothalamus, the hippocampus, the striatum and the gonadal tract.

In the spinal cord prodynorphin expression is markedly enhanced by nociceptive stimulation and by inflammatory and traumatic processes (Iadarola et al. 1988; Hollt et al. 1987). The increased prodynorphin expression provoked by noxious stimulation or inflammation is preceded by an early induction of c-*fos* in the same neurones (Naranjo et al. 1991; Noguchi et al. 1991).

The known sequence of the rat prodynorphin gene does not contain a classical AP-1 or CRE element. Transfection studies in which sequences of the 5'-flanking region of the rat prodynorphin gene were joined with a CAT reporter gene and introduced into 2RC Leydig cells revealed that a 210 bp fragment comprising 122 bp of the 5'-flanking sequences and 88 bp of exon 1 of the rat prodynorphin gene confers the response to cAMP. This genomic DNA fragment contains the nucleotide sequence element TAGCGTCAG (position 58 to 66) which is 80% homologous to the TGACGTCAG sequence present in many genes regulated by cAMP (McMurray et al. 1989). Interestingly, this element does not confer inducibility by the phorbol ester TPA indicating that AP-1 proteins might be not involved.

On the other hand, in cultured rat spinal cord cells the serotonin 1A receptor agonist 8-OH-DPAT (8-hydroxy-2-(di-n-propylamino)-tetralin) increases cAMP levels and enhances prodynorphin mRNA levels (Lucas et al. 1993). This effect, however, appears to involve the induction of AP-1 proteins rather than the activation of CREB or a similar constitutive transcription factor, since it could be blocked by incubating the cultured cells with c-*fos* antisense-oligonucleotides (Lucas et al. 1993).

Inspection of 5'-flanking sequences of the rat prodynorphin gene revealed the presence of several AP-1-like elements. One of these sequences (TGACAAACA, position -257/-249) appeared to constitute a noncanonical AP-1-element which was shown to be a target by Fos/Jun transactivation (Naranjo et al. 1991).

Recently, other CRE-like elements have been localized within upstream sequences of the rat prodynorphin gene (betweeen -1500 and -1600) which confer strong inducibility by cAMP (Dubner, personal communication).

In conclusion, although both CRE- and AP-1 like elements have been localized within the rat prodynorphin gene, the involvement of the respective transcription factors in the induction of the gene may be different in the various cell types and is largely unknown. In particular, a precise analysis by transactivation and anti-sense experiments is required to understand the regulation of this gene by transcription factors.

Proopiomelanocortin

Proopiomelanocortin (POMC) is the precursor protein for several peptides and hormones, such as adrenocorticotropic hormone, melanocyte stimulating hormones and β-endorphin. The gene is predominantly expressed in the pituitary, there being stimulated by hypothalamic corticotropin releasing factor (CRF) and repressed by glucocorticoids (Hollt 1983; Hollt 1993). Most of the research work dealing with regulation of the gene's transcription in the pituitary has been done on the rat and human POMC gene, part of which will be presented in this short overview.

It is well established that CRF stimulation of corticotroph cells leads to an elevation of intracellular cAMP (Reisine et al. 1985; Loeffler et al. 1986). The subsequent signal transduction pathway, however, is still a matter of investigation. An increase in the cAMP level activates protein kinase A (PKA) , which leads to a phosphorylation of numerous proteins (Edelman et al. 1987). A number of other genes whose transcription is responsive to cAMP possess an usually octameric palindromic promoter motif of the sequence TGACGTCA or at least the pentanucleotide core CGTCA ("CRE") (Montminy et al. 1986; Comb et al. 1986; Deutsch et al. 1988). Members of the CREB/ATF transcription factor family bind to this sequence (Montminy and Bilezikjian 1987; Hai et al. 1989; Hoeffler et al. 1990; Ruppert et al. 1992) and lead to an enhanced transcription after phosphorylation by PKA (Gonzalez and Montminy 1989). A very similar motif, TGACTCA, is the recognition site for AP-1 (Angel et al. 1987), consisting of heterodimers of the immediate early gene products Jun and Fos, or of Jun homodimers (Rauscher et al. 1988; Chiu et al. 1988). Due to the similarity of these motifs crosstalk between the two signal transduction pathways seems probable and indeed has been observed (Karin and Smeal 1992; Fink et al. 1991; Masquilier and Sassone Corsi 1992). To add to the complexity, the *fos* gene also displays on its promoter a CRE and thus becomes itself a target gene for the PKA/CREB pathway (Berkowitz et al. 1989).

Promoter sequences of the human and the rat POMC gene do not contain the classical CRE site. A single AP-1 site is present in the rat gene, located within the first exon at position +41. Boutillier et al. (1991; 1992) have presented evidence that CRF stimulation of AtT-20 cells, a corticotroph mouse pituitary cell line, leads to a Ca^{2+}/CaM kinase and PKA dependent increase in Fos, which is proposed to activate POMC gene expression via this AP-1 site. On the other hand, earlier studies by Roberts and coworkers (1987), working with transiently transfected AtT-20 cells, revealed that 5'-flanking sequences of the rat POMC gene spanning from nt -320 to nt -133 confer strong CRF inducibility upon the heterologous thymidine kinase (tk) promoter. Similar to these results Drouin et al. could demonstrate by gene transfer experiments in cultured cells and in transgenic mice that promoter sequences from nt -480 to nt +63 are sufficient for hormonal stimulation and tissue specific expression of the gene (Jeannotte et al. 1987; Tremblay et al. 1988; Drouin et al. 1989). In a more recent publication Liu et al. (1992) clearly showed that in transgenic mice the upstream promoter region between nt's -324 and -34 is necessary and sufficient for basal and induced POMC expression in the pituitary. Interestingly, within these sequences the AP-1 site at position +41 is excluded.

This is in good accordance with the results obtained from investigations performed on the human POMC gene. Here, Usui and coworkers (1989) presented evidence in transfection experiments that promoter sequences spanning from nt -417 to nt -97 mediate cAMP dependent stimulation of the POMC gene. Transfection experiments performed in this laboratory using human POMC-reporter gene constructs have revealed the ability of a central promoter region to mediate cAMP induction (Kraus and Hollt 1993), as well as high basal expression in human pituitary tissue (Kraus et al. 1993). Thus, employing a construct containing the 5'-flanking sequences from nt -415 to nt -223 high reporter gene activities were observed in primary cells derived from different human pituitary tumors, whereas reporter gene activities mesured in cells of other origine were significantly lower.

Several binding sites for potential transcription factors have been mapped using the DNaseI footprinting technique on the rat (Therrien and Drouin 1991) and human (Kraus et al. 1993) POMC gene promoter. Recently, Therrien and Drouin (1993) could show that a novel cell specific helix-loop-helix factor is required for pituitary expression of the gene. Another nuclear factor named "PO-B" has been mapped to bind to the region close to the TATA box between nt's -15 and -3 of the rat gene. It represents a novel transcription factor which is involved in basal gene transcription (Riegel et al. 1990; Wellstein et al. 1991). The group of Bishop (1990) has characterized a putative binding site for the transcription factor AP-2 within the mouse POMC gene promoter.

In recent studies performed in this laboratory, the cAMP-responsive region of the human gene could be delineated to promoter sequences spanning from nt -344 to nt -319. Reporter genes containing this promoter part were responsive to the cAMP elevating agent forskolin, and, an induction of reporter gene activity was also observed when the regulatory subunit of PKA alone, or together with CREB, was additionally coexpressed in AtT-20 cells. The DNA element corresponds to one of the previously defined protein binding sites (HPE-4) and consists of two octanucleotides, that are similar to the classical CRE sequence, spaced by three basepairs (Fig. 2).

Interestingly, the proenkephalin gene promoter also possesses two CRE sequences, ENKCRE-1 and ENKCRE-2, separated by five basepairs. They do not match to the palindromic CRE sequence and only ENKCRE-2 contains the core motif CGTCA.

Gel mobility shift assays in combination with antibodies indicate that CREB is at least one factor binding to this "POMC-CRE" (Fig. 3). The fact that we were unable to immunoshift a larger part of the DNA-protein complex may be due to the binding of additional factors or factors closely related to CREB, i.e. other factors of the CREB/ATF family (Kraus and Höllt, in preparation).

-344 CACGCAGG<u>TAACTTCA</u>CCC<u>TCGCCTCA</u>ACGACCT -319

Fig. 2. Nucleotide sequence of the human POMC CRE. The sequences which are homologous to the classical CRE (TGACGTCA) are underlined.

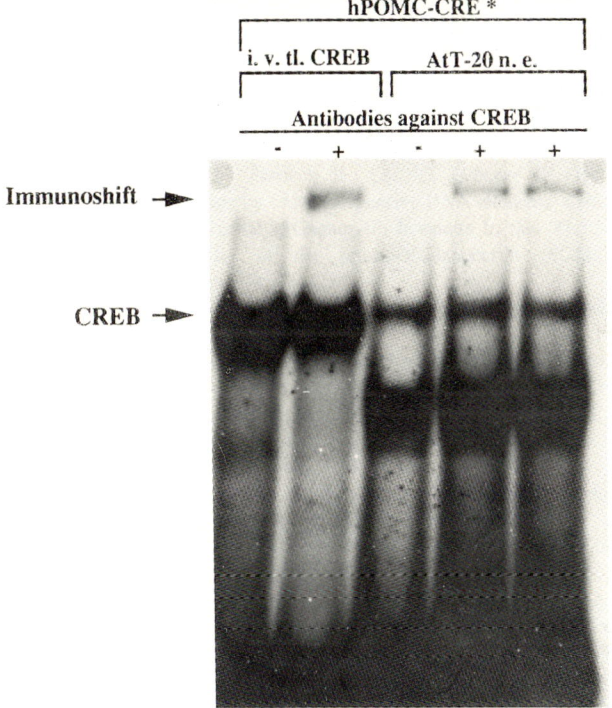

Fig. 3. AtT-20 nuclear proteins binding to the POMC CRE contain CREB. On a 5% native polyacrylamid gel 5 fmol per lane radioactive labelled POMC CRE (nt's -344 to -319 of the human gene) were subjected to electrophoresis after incubation with either *in vitro* translated CREB (lanes 1+2) or AtT-20 cell nuclear extract (lanes 3-5). Preincubation with CREB antibodies lead to additional ("supershifted") bands on top.

To conclude with, the rat POMC gene contains a classical AP-1 element and is transactivated by Fos, although this element seems unnecessary for POMC gene

Hyman SE, Comb M, Pearlberg J, Goodman HM (1989) An AP-2 element acts synergistically with the cyclic AMP- and phorbol ester-inducible enhancer of the human proenkephalin gene. Mol Cell Biol 9:321-324

Iadarola MJ, Douglass J, Civelli O, Naranjo JR (1988) Differential activation of spinal cord dynorphin and enkephalin neurons during hyperalgesia: evidence using cDNA hybridization. Brain Res 455:205-212

Jeannotte L, Trifiro MA, Plante RK, Chamberland M, Drouin J (1987) Tissue-specific activity of the pro-opiomelanocortin gene promoter. Mol Cell Biol 7:4058-4064

Karin M (1989) Complexities of gene regulation by cAMP. Trends Genet 5: 65-67

Karin M, Smeal T (1992) Control of transcription factors by signal transduction pathways: the beginning of the end. Trends Biochem Sci 17: 418-422

Kobierski LA, Chu HM, Tan Y, Comb MJ (1991) cAMP-dependent regulation of proenkephalin by Jun D and Jun B: positive and negative effects of AP-1 proteins. Proc Natl Acad Sci USA 88:10222-10226

Konradi C, Kobierski LA, Nguyen TV, Heckers S, Hyman SE (1993) The cAMP-response-element-binding protein interacts, but Fos protein does not interact, with the proenkephalin enhancer in rat striatum. Proc Natl Acad Sci USA 9:7005-7009

Kraus J, Buchfelder M, Hollt V (1993) Regulatory elements of the human proopiomelanocortin gene promoter. DNA Cell Biol 12:527-536

Kraus J, Hollt V (1993) Identification of hormone response elements of the human proopiomelanocortin gene promoter. Eur J Physiol 422:Suppl 1, R87. (Abstract)

Liu B, Hammer GD, Rubinstein M, Mortrud M, Low MJ (1992) Identification of DNA elements cooperatively activating proopiomelanocortin gene expression in the pituitary glands of transgenic mice. Mol Cell Biol 12: 3978-3990

Loeffler JP, Kley N, Pittius CW, Hollt V (1986) Calcium ion and cyclic adenosine 3',5'-monophosphate regulate proopiomelanocortin messenger ribonucleic acid levels in rat intermediate and anterior pituitary lobes. Endocrinology 119:2840-2847

Lucas JJ, Mellstrom B, Colado MI, Naranjo JR (1993) Molecular mechanisms of pain: serotonin1A receptor agonists trigger transactivation by c-*fos* of the prodynorphin gene in spinal cord neurons. Neuron 10:599-611

Masquilier D, Sassone Corsi P (1992) Transcriptional cross-talk: nuclear factors CREM and CREB bind to AP-1 sites and inhibit activation by Jun. J Biol Chem 267:22460-22466

McMurray CT, Devi L, Calavetta L, Douglass JO (1989) Regulated expression of the prodynorphin gene in the R2C Leydig tumor cell line. Endocrinology 124:49-59

Montminy MR, Sevarino KA, Wagner JA, Mandel G, Goodman RH (1986) Identification of a cyclic-AMP-responsive element within the rat somatostatin gene. Proc Natl Acad Sci USA 83:6682-6686

Montminy MR, Bilezikjian LM (1987) Binding of a nuclear protein to the cyclic-AMP response element of the somatostatin gene. Nature 328:175-178

Morgan JI, Curran T (1991) Stimulus-transcription coupling in the nervous system: involvement of the inducible proto-oncogenes *fos* and *jun*. Annu Rev Neurosci 14:421-451

Naranjo JR, Mellstrom B, Achaval M, Sassone Corsi P (1991) Molecular pathways of pain: Fos/Jun-mediated activation of a noncanonical AP-1 site in the prodynorphin gene. Neuron 6:607-617

Nishimori T, Buzzi MG, Moskowitz MA, Uhl GR (1989) Preproenkephalin mRNA expression in nucleus caudalis neurons is enhanced by trigeminal stimulation. Brain Res Mol Brain Res 6:203-210

Noguchi K, Kowalski K, Traub R, Solodkin A, Iadarola MJ, Ruda MA (1991) Dynorphin expression and Fos-like immunoreactivity following inflammation induced hyperalgesia are colocalized in spinal cord neurons. Brain Res Mol Brain Res 10:227-233

Rauscher FJ, Voulalas PJ, Franza BRJ, Curran T (1988) Fos and Jun bind cooperatively to the AP-1 site: reconstitution *in vitro*. Genes Dev 2:1687-1699

Reisine T, Rougon G, Barbet J, Affolter HU (1985) Corticotropin-releasing factor-induced adrenocorticotropin hormone release and synthesis is blocked by incorporation of the inhibitor of cyclic AMP-dependent protein kinase into anterior pituitary tumor cells by liposomes. Proc Natl Acad Sci USA 82:8261-8265

Riegel AT, Remenick J, Wolford RG, Berard DS, Hager GL (1990) A novel transcriptional activator

(PO-B) binds between the TATA box and cap site of the pro-opiomelanocortin gene. Nucleic Acids Res 18:4513-4521

Roberts JL, Lundblad JR, Eberwine JH, Fremeau RT, Salton SR, Blum M (1987) Hormonal regulation of POMC gene expression in pituitary. Ann N Y Acad Sci 512:275-285

Ruppert S, Cole TJ, Boshart M, Schmid E, Schutz G (1992) Multiple mRNA isoforms of the transcription activator protein CREB: generation by alternative splicing and specific expression in primary spermatocytes. EMBO J 11:1503-1512

Sonnenberg JL, Rauscher FJ, Morgan JI, Curran T (1989) Regulation of proenkephalin by Fos and Jun. Science 246:1622-1625

Stachowiak MK, Hong JS, Viveros OH (1990) Coordinate and differential regulation of phenylethanolamine N-methyltransferase, tyrosine hydroxylase and proenkephalin mRNAs by neural and hormonal mechanisms in cultured bovine adrenal medullary cells. Brain Res 510: 277-288

Sukhatme VP, Cao XM, Chang LC, Tsai Morris CH, Stamenkovich D, Ferreira PC, Cohen DR, Edwards SA, Shows TB, Curran T et al (1988) A zinc finger-encoding gene coregulated with c-*fos* during growth and differentiation, and after cellular depolarization. Cell 53:37-43

Therrien M, Drouin J (1991) Pituitary pro-opiomelanocortin gene expression requires synergistic interactions of several regulatory elements. Mol Cell Biol 11:3492-3503

Therrien M, Drouin J (1993) Cell-specific helix-loop-helix factor required for pituitary expression of the pro-opiomelanocortin gene. Mol Cell Biol 13:2342-2353

Tremblay Y, Tretjakoff I, Peterson A, Antakly T, Zhang CX, Drouin J (1988) Pituitary-specific expression and glucocorticoid regulation of a proopiomelanocortin fusion gene in transgenic mice. Proc Natl Acad Sci USA 85:8890-8894

Usui T, Nakai Y, Tsukada T, Fukata J, Nakaishi S, Naitoh Y, Imura H (1989) Cyclic AMP-responsive region of the human proopiomelanocortin (POMC) gene. Mol Cell Endocrinol 62:141-146

Van Nguyen T, Kobierski L, Comb M, Hyman SE (1990) The effect of depolarization on expression of the human proenkephalin gene is synergistic with cAMP and dependent upon a cAMP-inducible enhancer. J Neurosci 10:2825-2833

Wang X, Bacher B, Hollt V (1993) Gene expression in bovine adrenal chromaffin cells: relationship between c-Fos/c-Jun and proenkephalin. Eur J Physiol 422, Suppl 1, R92. (Abstract)

Wellstein A, Dobrenski AF, Radonovich MN, Brady JF, Riegel AT (1991) Purification of PO-B, a protein that has increased affinity for the pro-opiomelanocortin gene promoter after dephosphorylation. J Biol Chem 266:12234-12241

Transgenic mice studies of immediate-early genes: from markers to mutants

G. Kasof [1,2]*, T. Curran*[1]*, and J.I. Morgan*[1]

[1]Roche Institute of Molecular Biology, Roche Research Center Nutley, NJ 07710
[2]Dept. of Biological Sciences, Columbia University, New York, NY 10027

Introduction

Eurokaryotic cells can selectively modify gene transcription in response to alterations in the levels of extracellular signals. In general, genomic targets of extracellular stimuli can be devided into two categories based upon the characteristics of their regulation (reviewed in Curran and Morgan 1987; Morgan and Curran 1989). Thus, alterations in transcription of "primary response" genes is independent of protein synthesis and is typically initiated within minutes. In contrast, changes in the expression of "delayed", or "late response" genes occur with a lag of several hours and requires ongoing protein synthesis. This nomenclature is based partly upon analogies with certain viruses that control their replicative cycle by the sequential activation of groups of genes (i.e. immediate-early, early and late genes). As their name suggests, viral immediate-early genes are transcriptionally active immediately upon infection of the host cell and are transcribed even in the absence of protein synthesis. Their products serve to coordinate further progression of the viral life cycle by regulating transcription of viral early genes (reviewed in Berk 1986). This has led, again by analogy, to the suggestion that the proteins encoded by the cellular primary response genes contribute to the transcriptional regulation of late response genes (Goelet et al. 1986; Morgan and Curran 1989). This would also imply that primary response gene products should encode inducible transcription factors or proteins capable of modifying the activity of existing transcription factors.

Several strategies have evolved for initiating the primary genomic response. First, membrane-permeant stimuli, such as steroids, retinoids, and thyroid hormone, can alter transcription by binding directly to preexisting nuclear receptors. Most of these receptors belong to a now well-characterized superfamily of transcription factors. In a second situation, extracellular stimuli trigger a cascade of intracellular signalling events that culminate in the activation of primary response gene transcription. Such intracellular signalling events usually involve as a final step the post-translational modification of existing transcription factors. For example, following cell stimulation several kinases are known to phosphorylate proteins of the CREB (cAMP response element binding protein) or ATF (activating transcription factor) families of transcription factors. These proteins are thought to be bound to their DNA target and it is their transcriptional activity that is altered by phosphorylation. Alternatively, the signalling cascade may induce the translocation of a transcription factor from an extranuclear, or inactive, site to the nucleus (e.g. NF-κB). There is now a considerable interest in the mechanisms involved in stimulus-transcription coupling and this will be reviewed here from the perspective of immediate-early gene expression in the nervous system.

Cellular immediate-early genes (IEGs) are included amongst those genes that comprise the primary genomic response. Their expression is generally low in most cells, but they are induced in a rapid, and frequently transient, manner by a wide variety of stimuli including neurotransmitters, growth factors, and hormones (reviewed in Curran and Morgan 1987; Hershman 1989). Like viral IEGs this induction is independent of protein synthesis and generally is activated via an intracellular signalling cascade. Since many IEGs were first identified in the context of cell proliferation and cellular transformation (several are known proto-oncogenes) there was a misconception that their function was solely to regulate the cell cycle. However, many studies, particularly those involving cells from the nervous system, have shown that they function in diverse biological contexts unrelated to proliferation, for example, in differentiation, regeneration, neuronal excitation and even cell death.

Although the IEG class is believed to contain as many as one hundred genes, relatively few have been characterized to any great extent. The genes, c-*fos* and c-*jun*, have served as the prototypes of IEG. c-*fos* and c-*jun* are the cellular homologues of the transforming genes isolated from the FBJ murine osteosarcoma virus (v-*fos*) and avian sarcoma virus-17 (v-*jun*), respectively (Curran and Teich 1982; Maki et al. 1987). Their genes encode nuclear phosphoproteins, Fos and Jun, that contain a leucine zipper motif and basic DNA binding domain (Landschulz et al. 1988; Turner and Tjian 1989; Schuermann et al. 1989; Gentz et al. 1989; O'Shea et al. 1992). Fos and Jun form heterodimers capable of binding to, and regulating transcription from, AP-1 regulatory sites (Bohmann et al. 1987; Angel et al. 1988; Rauscher et al. 1988). In addition, it has been shown that Jun (but not Fos) can form homodimers that possess transcriptional activity (Nakabeppu et al. 1988). Furthermore, several genes that are structurally related to c-*fos* and c-*jun* have been discovered. *fra*-1 (Cohen and Curran 1988), *fra*-2 (Matsui et al. 1990) and *fos* B (Zerial et al. 1989) comprise the *fos* family; while *jun* B (Ryder et al. 1988) and *jun* D (Ryder et al. 1989) are closely related to c-*jun*. Members of these families share the leucine zipper motif and DNA binding domain and belong to the basic-zipper superfamily of transcription factors, which also includes CREBs and ATFs. Many members of this superfamily are capable of homo- and heterodimer formation, thereby providing the potential for great diversity in terms of dimerization and DNA binding specificities. For example, Fos-Jun dimers bind preferentially to AP-1 sites while CREBP1-Jun dimers interact preferentially with CRE sites (Hai and Curran 1991). Taken together, one should view the specificity of this response as being dictated by the complement of inducible and constitutive transcription factors present in a given cell following application of a stimulus.

Other members of the IEG class encode proteins with structural motifs characteristic of transcription factors. These include c-Myc, a helix-loop-helix protein (Murre et al. 1989); NGFI-A (*zif*/268, *egr* 1 or *krox* 24), a zinc finger protein (Milbrandt 1987; Sukhatame et al. 1988); and NGFI-B (*nur*/77), a steroid receptor class of protein (Hazel et al. 1988). However, it should be emphasized that several known IEGs encode non-nuclear proteins, such as cytokines and phosphatases. It is believed that these gene products serve to alter stimulus-response coupling and to transmit the immediate-early response to surrounding cells. It is this realization that has lead to the more global view of a cellular immediate-early response which coordinates and propagates an integrated biological program (Hilbush et al. 1994).

Using IEGs for metabolic mapping

Because of their inducibility in neurons, the detection of IEG products (particularly Fos) has been used as a surrogate form of metabolic mapping in the nervous system. Since Fos is localized to the nucleus, its detection offers a level of cellular resolution not obtainable with 2-deoxyglucose (2-DG) autoradiography. While this method has a number of limitations, it does provide another window through which neurophysiological processes can be viewed *in vivo*. The principal concerns with this technique are (1) that classical neuronal excitation does not necessarily result in c-*fos* in all circumstances and (2) that instances are known in which the gene is induced although there is no enhanced neuronal activity. For example, PTZ induces 2-DG uptake in the substantia nigra and cerebellum but does not increase Fos-like immunoreactivity (Fos-LI) in these regions (Ben-Ari et al. 1981; Morgan et al. 1987). Following destruction of dopaminergic neurons with 6-hydroxydopamine (6-OHDA), injection of L-DOPA increased 2-DG uptake in substantia nigra without increasing Fos-LI (Trugman and Wooten 1986; Robertson and Robertson 1987; Robertson et al. 1989). In contrast, Fos-LI has been observed in the paraventricular and supraoptic nuclei following 24 hr. water deprivation although no increase in 2-DG uptake was detected (Sagar et al. 1988). Finally, Jorgenson et al. (1989) have shown that cerebral ischemia induces c-*fos* in hippocampal area CA1 without an associated increase in 2-DG uptake. Thus, Fos immunohistochemistry and 2-DG maps are not always coincident. In many ways this is to be expected since c-*fos* transcription is sensitive to alterations in many second messenger signalling cascades while 2-DG uptake is a measure of metabolic activity. Clearly, second messenger levels can be modified by many types of stimuli that do not necessarily result in altered neuronal firing and energy consumption. On the other hand, 2-DG uptake is a graded response, the more firing, the greater the energy consumption. In contrast, the recruitment of the IEG response is a threshold phenomenon (Shin et al. 1990), if the threshold is not met, there is no induction despite the fact that firing has increased.

Despite the above concerns and other caveats associated with the technique, IEG expression has been followed in a number of neuronal activity models. From these experiments the following general points have been established that should be used when designing and interpreting experiments using detection of IEG products in the nervous system. (1) As noted above, the absence of Fos cannot be interpreted absolutely as meaning that a particular neuron did not fire in a particular situation. Likewise, the presence of Fos does not indicate absolutely that a neuron increased firing. (2) It should be recognized that there is basal expression of Fos-LI in the brain, particularly in the pyriform cortex, olfactory nucleus, and bed nucleus of stria terminalis (Morgan et al. 1987). (3) Stress, elicited by injection of saline or by immobilization, induces Fos-LI in the paraventricular nucleus of the hypothalamus and the intermediate and anterior lobes of the pituitary (Kononen et al. 1992). Thus, care should be taken in how animals are handled. (4) The level of Fos-LI exhibits a circadian rhythm in some brain regions (Kononen et al. 1990), making consistent timing of the experiments critical. (5) Biochemical analyses have established that waves of Fos-like proteins appear and disappear in the brain in a staggered manner. Many of these proteins cross-react with commercially available Fos antisera. Thus, the time lag between applying a stimulus and sacrificing the animal as well as the details of the specificity of the antiserum are crucial for experimental design and interpretation (what proteins are being detected). (6) Many anesthetics interfere with IEG expression, therefore, experiments which require anesthesia are particularly difficult to interpret in the absence of appropriate controls.

The usefulness of IEG expression as a marker for synaptic activity *in vivo* has been examined extensively (reviewed in Morgan and Curran 1991a). Acute treatment with the convulsant drug, pentylenetetrazole (PTZ) produced a massive and transient induction of c-*fos* mRNA and Fos-LI in the pyriform and cingulate cortices and the hippocampus (Morgan et al. 1987). c-*fos* induction by PTZ is blocked by anticonvulsants such as diazepam and sodium pentobarbital, suggesting that the seizure is necessary for Fos induction. Other IEGs are co-induced by PTZ, including c-*jun*, *jun* B and *zif*/268 (Saffen et al. 1988; Sonnenberg et al. 1989a). In addition to PTZ, the glutamate receptor agonists, NMDA (N-methyl-D-aspartate) and kainic acid, also induce seizures that are associated with increases in IEG expression (Sonnenberg et al. 1989a). While the spectrum of genes induced by the various convulsants is somewhat stereotyped, there are some subtle differences in their kinetics and quite marked differences in their brain distribution and pharmacology. For example, kainic acid induced c-*fos* expression is higher and more protracted than that seen with either NMDA or PTZ (Sonnenberg et al. 1989a; Molinar-Rode et al. 1993). However, PTZ produces a more robust induction of Fos-LI in cortex than kainic acid while the converse is true in the limbic system. These differences are thought to be related to the different distributions of various glutamate receptors. Kainic acid seems to act via the kainate/AMPA (GluR) type of glutamate receptor which is known to be concentrated in the hippocampus. Based upon its pharmacology (blocked by MK-801), PTZ appears to induce *fos* by way of NMDA (NMDAR) type glutamate receptors. Furthermore, the distribution of NMDA receptors is largely coincident with PTZ-induced Fos-LI in the CNS (Morgan et al. 1987; Sonnenberg et al. 1989a). While an elevation of intracellular Ca^{++} is required for NMDA- and kainic acid-induced increases in c-Fos, different second messenger pathways must be involved (Lerea and McNamara 1993). Indeed, Bading et al. (1993) have shown that Ca^{++} channels activate distinct intracellular pathways and stimulate transcription of c-*fos* by means of different *cis*-acting regulatory elements.

In addition to acute chemoconvulsant paradigms, the regulation and function of

IEGs have been addressed in neuronal models that involve long-term changes in synaptic activity. Kindling is a model for human complex partial epilepsy that appears to require protein synthesis for its establishment and that is mediated by NMDA receptors. In this model, repeated application of an initially subconvulsive electrical stimulus results, eventually, in generalized clonic motor seizures (Goddard et al. 1969). Electrical stimulation of the angular bundle at levels that produce kindling also lead to a transient induction of *fos* mRNA and Fos-LI in the dentate gyrus and areas CA1 and CA3 of the hippocampus (Dragunow and Robertson 1987; Shin et al. 1990). However, there is no difference between the kinetics and magnitude of *fos* induction in the hippocampi of naive and kindled rats (Shin et al. 1990). Therefore, aberrant Fos expression is unlikely to contribute to the kindled state *per se*, although Fos could play a role in the establishment of kindling. The kindled state is associated with changes in expression of several neuromodulators including proenkephalin (McNamara et al. 1987) as well as the physical sprouting of axons in the hippocampus (Sutula et al. 1988). Thus it has been proposed that IEG expression could be contributing to these processes (Morgan and Curran 1991b).

In the PTZ seizure paradigm it is known that the induction of c-*fos*, c-*jun* and *jun* B in the hippocampus precedes an increase in proenkephalin mRNA (Kanamatsu et al. 1986; White and Gall 1987; Sonnenberg et al. 1989a, 1989b). Furthermore, the proenkephalin promoter contains a potential AP-1 binding site and it can be activated by Fos and Jun and other members of the basic-zipper gene family in transient transfection assays (Sonnenberg et al. 1989b; Kobierski et al. 1991). Thus there is plausible evidence to suggest that inducible and constitutive members of the basic-zipper superfamily may contribute to the up-regulation of proenkephalin in the hippocampus in both acute and chronic seizure models. In addition to proenkephalin, NGF and BDNF mRNAs have been shown to increase in the hippocampus following electrical stimulation and/or seizures (Isackson et al. 1991; Patterson et al. 1992; reviewed in Morgan and Curran 1991b). Indeed, it has been proposed that NGF transcription is driven via an AP-1 site (Hengerer et al. 1990). Thus a scenario has been proposed in which electrical activity induces an IEG response which results in the expression of elevated levels of neurotrophins that in turn elicits a growth response manifested as axonal sprouting (Morgan and Curran 1991b).

Cerebral ischemia can cause neuronal dysfunction, structural damage and cell death. However, ischemia is also associated with changes in gene expression (Jacewitz et al. 1986; Dempsey et al. 1991; Welsh et al. 1992) that could be mediated by IEG products. Indeed, brain injury has been shown to induce Fos-LI in nerve and glial cells (Dragunow and Robertson 1988) and both global and focal cerebral ischemia have been reported to variously affect expression of c-*fos*, *jun* B, and to a lesser degree, c-*jun* (An et al. 1993). Recently, it has been established that hippocampal and amygdaloid neurons that undergo delayed cell death as a result of kainic acid neurotoxicity also express IEGs (Smeyne et al. 1993). However, in these situations it is difficult to establish cause and effect. Is Fos neuroprotective? Does it contribute to cell death? Is its induction an epiphenomenon triggered by a collapse of intracellular signalling cascades? Alternative strategies will be required to resolve this type of issue and the balance of this text will discuss methods for resolving this and other questions regarding the role of IEG products in the nervous system.

Fos-lacZ transgenic rodents

To follow the expression of Fos unambiguously and quickly we have taken advantage of transgenic rodent technology. Recently, transgenic mice were bred that carry a *fos-lacZ* /ß-galactosidase3) coding sequence fused in frame to exon 4 of c-*fos* (Fig. 1; Schilling et al. 1991). This gene contains all of the known 5'-upstream regulatory sequences of c-*fos* as well as all introns and exons and several kilobases of 3' flanking genomic DNA. The mRNA generated by this gene contains in its 3'-untranslated region sequences that confer a short half-life to the c-*fos* mRNA. Therefore, this *fos-lacZ* gene should not only show identical kinetics of induction as c-*fos* but its mRNA should be rapidly degraded like c-*fos* mRNA. The Fos-lacZ fusion protein generated from this gene contains the nuclear localization sequences from Fos that result in ß-galactosidase being translocated to the nucleus. Here it can be readily detected by histochemistry with single cell resolution.

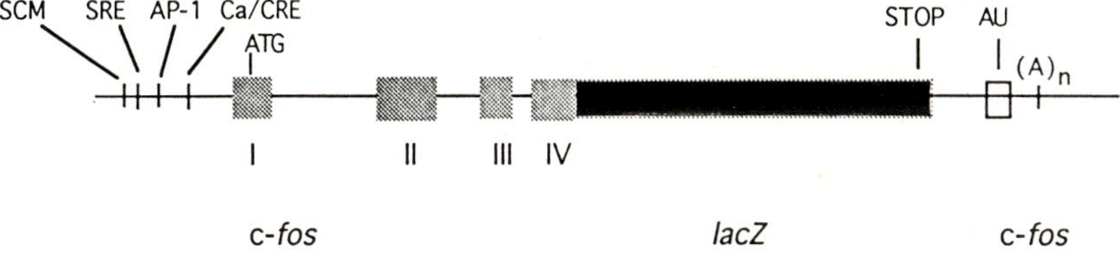

Fig. 1. Structure of the *fos-lacZ* fusion gene. Several features of the fusion gene are illustrated. The translational start and stop sites are indicated by the labels, ATG and STOP, respectively. In the 5' untranslated region of c-*fos*, the transcriptional regulatory sequences, including the *sis*-conditioned medium (SCM) element, serum response element (SRE), AP-1 binding site, and the calcium/cAMP response element (Ca/CRE) are labeled. Exons of c-*fos* and *lacZ* are illustrated by the hatched and solid boxes respectively. The 3' untranslated region contains the A + U-rich region (AU) and polyadenylation signal [(A)$_n$] of c-*fos*.

Using the Fos-lacZ transgenic model we first established that the expression of the fusion gene accurately recapitulated the distribution of Fos-LI following PTZ and kainic acid seizures (Fig. 2; Smeyne et al. 1992a). Subsequently, we have investigated the development of the brain in terms of the onset of synaptic activity. For example, perinatal mice show a characteristic distribution of Fos-lacZ in several regions of the CNS; notably cingulate gyrus, distal CA1 of hippocampus, parafascicular nucleus of the thalamus and mitral cell of the olfactory bulb. Typically this expression disappears during the first week of life (Smeyne et al. 1992b). It was argued that this expression might be related to plasticity and establishment of functional circuitry in the postnatal nervous system. Presently, experi-

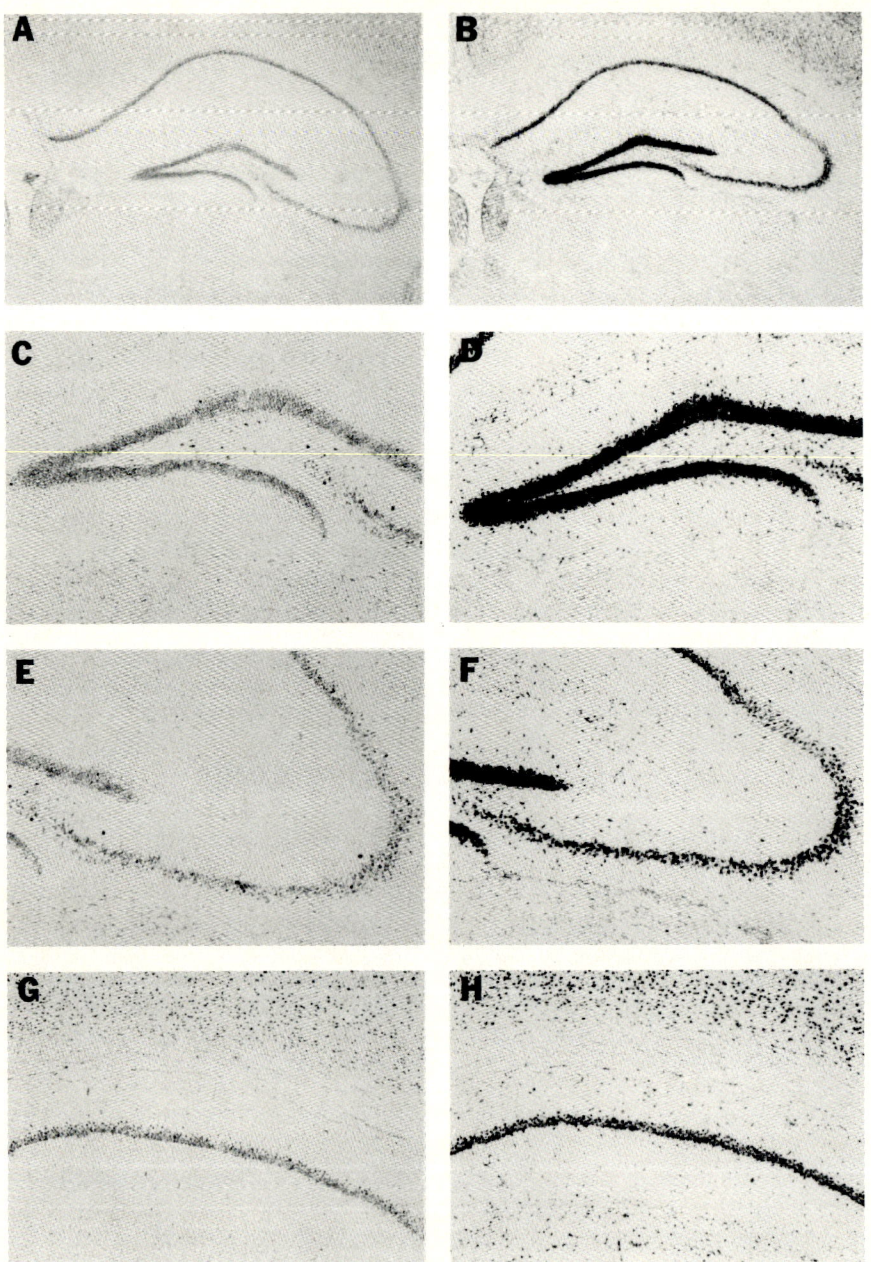

Fig. 2. Expression of Fos-lacZ in the hippocampus following seizures. Seizures were induced by intraperitoneal injection of kainic acid (KA; 17mg/kg) in transgenic mice that carry the *fos-lacZ* fusion gene. Preparation of tissues and staining with X-gal (5-bromo-4-chloro-3-indolyl-ß-D-galactopyranosidase) have been described previously (Smeyne et al. 1992a). Shown above are the hippocampi from a control, unseizured mouse (A, C, E, G) and a mouse injected with KA two hours prior to fixation (B, D, F, H). Appearance of blue nuclei indicate the presence of Fos-lacZ. Higher magnification images are shown of the dentate gyrus (C and D), and hippocampal areas CA3 (E and F) and CA1 (G and H).

ments are being performed to relate the inducibility of the *fos-lacZ* transgene in particular neural substructures to the stage of their maturation. For example, in one model we have established that the appearance of one class of glutamate receptor type is associated with the acquisition of IEG inducibility in the hippocampus (Vendrell et al., unpublished).

The localization of expression of IEG products will provide not only evidence of synaptic activity but also clues to the function of these genes. A recent surprising finding was that continuous expression of the *fos-lacZ* transgene is associated with cells undergoing terminal differentiation resulting in (programmed) cell death (Smeyne et al. 1993). Naturally occuring examples of this include the cells within the skin and hair follicle, interdigital web cells, follicular cells within atretic follicles in the ovary and hypertrophic chondrocytes. In terms of the nervous system we have determined that neurotoxic doses of kainic acid induce the gene in areas CA1 and CA3 of the hippocampus as well as the amygdala. Interestingly, this induction occurs with a delay of four days and has the unusual property that the Fos-lacZ is no longer confined to the nucleus. This is probably attributable to the imminent demise of the cell which may be associated with a breakdown of nuclear integrity. In parallel studies, a transgenic rat carrying the same *fos-lacZ* transgene yielded a qualitatively similar result following administration of kainic acid (Kasof et al., unpublished).

Programmed cell death is also used extensively in the developing nervous system to match the size of one neuronal population with its target. This often involves competition between neurons for target-derived trophic factors during a critical time period of development. In one such model, neonatal transgenic mice received a unilateral transection of the sciatic nerve (mimicking deprivation of target-derived growth factors) which results in the death of some motor neurons in the ventral horn of the spinal cord and cells in the dorsal root ganglion. As in the kainic acid lesion, Fos-lacZ was expressed in populations of cells that die in this model (Smeyne et al. 1993).

A powerful extension to the transgenic mouse strategy that needs to be pursued further is to cross these animals to one of the many strains of rodents that carry mutations that cause neuropathological conditions. For example, in an attempt to find further evidence for conditions in which Fos expression is associated with cell death the *fos-lacZ* mouse was mated with the *weaver* mutant mouse (Smeyne et al. 1993). This mouse strain is known to exhibit cell death in the granule layer of the cerebellum during the immediate postnatal period. As in the other models, Fos-lacZ was expressed in cells located in the position of degenerating granule neurons in *fos-lacZ* carrying *weaver* mice.

The *fos-lacZ* model can be extended to investigate other aspects of neurophysiology and IEG expression. For example, the promoter of c-*fos* contains a number of regulatory elements that confer inducibility by distinct signalling pathways (reviewed in Morgan and Curran 1991a). For example, calcium and cyclic AMP control c-*fos* expression by way of the CaRE (Ca/cAMP response element) site, whereas NGF appears to exert its action via a distinct site, the SRE (serum response element). By introducing point mutations into known response elements of the c-*fos* promoter one can analyze those trans-acting factors necessary for gene induction by specific stimuli *in vivo*. Using this strategy, Greenberg and colleagues have determined that a CREB protein binds at the CaRE site and upon phosphorylation by the cyclic AMP-dependent protein kinase or the calcium-calmodulin dependent protein kinase II activates transcription of c-*fos* (Sheng et al. 1990). Traditionally these type of promoter analyses have been performed by transient transfection (Prywes et al. 1998; Sheng et al. 1990) and the reporter constructs typically lack 3' and intragenic sequences that may be relevant to gene regulation (Lamb et al. 1990). Also, one is restricted to particular cell types and the approach does not provide for the analysis of complex biological responses *in vivo*. However, the transgenic approach provides a route for circumventing these shortcomings. Therefore, we have begun to analyze the expression of Fos-lacZ in mice harboring point mutations of some of its known regulatory elements including the SRE and CaRE sites. The hope is that this information will provide clues to the signal transduction pathways that are active in particular cells *in vivo* following administration of specific stimuli. In addition, it will provide evidence for the contribution of particular responsive elements to c-*fos* expression *in vivo*. Our initial analysis indicates that signalling to the c-*fos* promoter *in vivo* is more complex than in cell culture. For example, while only a CaRE element is required for induction of c-*fos* by depolarizing stimuli in PC12 cells, several elements are necessary for *fos-lacZ* induction by seizures. Recent evidence from hippocampal neurons in culture also indicates that other promoter elements may be involved in mediating the induction of c-*fos* via the NMDA receptor (Bading et al. 1993).

Function of IEGs

The analysis of IEG function *in vivo* is much more problematical and, as for proenkephalin, prodynorphin and NGF regulation, relies heavily upon indirect evidence such as spatio-temporal coincidence of gene expression (Draisci and Iadarola 1989). In some instances the case has been strengthened by transient transfection assays in cell culture (Sonnenberg et al. 1989b; Kobierski et al. 1991) and gel retardation studies using control and stimulated brain extracts (Sonnenberg et al. 1989c; Konradi et al. 1993). However, while this has pointed to important aspects of gene regulation, it does not provide direct proof of causation *in vivo*. Indeed, it is perhaps somewhat narrow to view gene regulation in terms of whether a single transcription factor uniquely regulates a particular target gene. It is likely that the regulation of any particular gene results from combinatorial interactions among many transcription factors as well as their state of posttranslational modification. Nevertheless, a number of approaches are available that permit us to begin unravelling the complex issue of IEG involvement in biological processes.

One approach to studying IEG function is to limit the amount of a particular IEG product and measure the effect this has on a particular phenotype. While this approach is not practicable *in vivo* it can be used in culture or, potentially, brain slices. Antibodies against c-Fos have been injected into fibroblasts where they have been found to disrupt the cell cycle. In fact, Fos antibodies blocked serum stimulated DNA synthesis in quiescent (serum-starved) fibroblasts even when injected up to 6 to 8 hours after serum readdition (Riabowol et al. 1988).

An alternative method of reducing the level of an IEG product is by the use of antisense oligonucleotides which has gained popularity in recent years. Blocking the expression of IEGs with antisense oligonucleotides has been used somewhat more successfully than antisera in efforts to determine target genes as well as the role if IEGs in some physiological processes. The most striking experiments of this type have usually employed simultaneous application of oligonucleotides that are antisense to both c-*fos* and c *jun*. The types of systems studied have usually involved ligand stimulated increase in putative target genes or the secretion of their products. For example, serotonin receptor 1a agonists induce preprodynorphin expression in spinal cord cultures. This induction can be blocked by c-*fos* and c-*jun* antisense (Lucas et al. 1993). c-*fos* antisense oligonucleotides injected into PC12 cells blocked NGF induction of Fos by about 50%, but was not effective in preventing neurite outgrowth (Kindy and Verma 1988). This suggested that Fos was not essential for the neuritogenic effects of NGF. In another case, antisense oligonucleotides to c-*fos* and c-*jun* blocked interleukin-1(IL-1)-induced ß-endorphin secretion in the mouse pituitary cell line, AtT-20 (Fagarasan et al. 1990). Antisense experiments have also been used to probe overall biological consequences. For instance, programmed cell death of lymphoid cells induced by growth factor deprivation was also blocked by a combination of c-*fos* and c-*jun* antisense oligonucleotides (Colotta et al. 1992). Recently, several investigators have studied the effects of antisense constructs on behavioral responses in rodents. Direct infusion of antisense oligonucleotides to c-*fos* into striatum reduced amphetamine-induced production of Fos-LI (Chaisson et al. 1992).

While antisense technology does provide a relatively rapid route to alter the levels of IEG products *in vitro* it is by no means a well established method in living animals and there are some caveats with its use (reviewed in Akhtar and Juliano 1992). Therefore, to understand better the function of IEGs *in vivo*, investigators have taken advantage of transgenic models. In the case of Fos and the AP-1 complex, five approaches are possible: (1) over-expressing the gene product on a heterologous promoter; (2) reducing the levels of the protein by driving an antisense construct *in vivo*; (3) inactivating the gene by homologous recombination; (4) introducing *trans*-dominant mutations that sequester the IEG product; (5) introducing mutant forms of the IEG product that bind to AP-1 sites but do not affect transcription.

In an attempt to disrupt expression of c-*fos*, embryonic stem cells were transfected with DNA in which exon 2 of c-*fos* was replaced with a neomycin resistant gene. By homologous recombination this yielded ES cells carrying an inactivated c-*fos* gene (Wang et al. 1992; Johnson et al. 1992). These cells were then microinjected into the blastocysts from C57BL/6 mice and mice bred to obtain germline transformants. Transgenic mice heterozygous for the c-*fos* null mutation appeared normal. However, homozygous mutants displayed retarded growth, deficiencies in bone remodeling and tooth eruption, perturbed folliculogenesis, and altered hematopoiesis. No effect was observed on fibroblast cell cycle as had been demonstrated *in vitro* using antibodies to interfere with c-*fos* function. However, since *fos* belongs to a gene family it is not clear whether compensation can occur in AP-1 function by use of alternative heterodimer configurations. Certainly, redundancy amongst tyrosine kinases results in very modest phenotypes when they are knocked-out individually. To try to circumvent redundancy, attempts have been made to antagonize AP-1 function rather than eliminate individual components. This approach is also favored since mice null for some other IEGs, including c-*jun* (Hilberg and Wagner 1992) and N-*myc* (Sawai et al. 1991), are embryonic lethals and so their contribution to adult neurophysiology cannot be determined.

The trans-dominant mutation approach has been a powerful technique in blocking function of transcription factors which bind DNA in a dimeric complex. Mutations have been introduced into the coding regions of *fos* and *jun* that eliminate DNA binding activity (i.e. mutations of the basic domain) but preserve the protein-protein association motif (leucine zipper). Thus, mutant protein can dimerize with native partners, but they no longer bind DNA and are, therefore, inactive. Such investigations have been performed primarily *in vitro*. When expressed continuously at high levels, c-*fos* and c-*jun* can cause transformation (Miller et al. 1984; Schuermann et al. 1989). However, cells transfected with a mutated c-*fos* gene lacking the DNA-binding domain renders cells resistant to transformation by c-*jun* and also reverts c-*jun* transformed cells back to their normal phenotype (Okuno et al. 1991). This reversion of transformation has been associated with the appearance of mutant Fos-Jun heterodimers as well as a reduction in AP-1 DNA-binding activity. Other trans-dominant mutations that have been analyzed include *jun* (Lloyd et al. 1991) and the rel/NF-kB family (Logeat et al. 1991).

In addition to inactivating a gene, transgenic mouse technology permits regulated gene expression either globally or in a tissue specific manner. For example, c-*fos* has been linked to the human metallothionein promoter and introduced into transgenic mice (Ruther et al. 1987). Thus expression of Fos can be induced by heavy metals which activate the metallothionein promotor. Expression of the transgene was seen in many tissues with highest levels in the pancreas, kidney, brain, heart, and muscle. Furthermore, Fos levels could be increased ten-fold by injection of $CdCl_2$. As with the *fos*-null animals, these mice displayed abnormal bone formation. In another case, c-*myc* was linked to the mammary tumor virus (MTV) promoter resulting in mammary adenocarcomas (Stewart et al. 1984). Studies are now aimed at utilizing the growing number of tissue specific promoters, especially as several of these can direct gene expression in subpopulations of neurons. For example, the olfactory marker protein (OMP) gene, which is expressed uniquely in olfactory receptor neurons, has been used to specifically direct expression of the SV40 large T antigen to these cells (Largent et al. 1993). Also, the L7 promoter can direct heterologous gene expression to cerebellar Purkinje cells and retinal rod bipolar neurons (Oberdick et al. 1990). Since these promoters can be used to drive mutant and antisense constructs as well as cognate sequences, this approach offers great potential and will allow gene manipulation in defined neurons *in vivo*.

Conclusions

Despite the enormous attention IEGs have received during the past decade, their function in neuronal physiology remains controversial. However, investigators have begun to move from the phenomenological aspects of IEG research (i.e. determining which IEGs are induced in what cells following a particular stimulus) to studies addressing the function that they subserve. The application of some of the technologies and strategies outlined here may be useful in this quest particularly as it pertains to the nervous system.

References

Akhtar S, Juliano RL (1992) Cellular uptake and intracellular fate of antisense oligonucleotides. Trends Cell Biol 2:139-144

An G, Lin TN, Liu JS, Xue JJ, He YY, Hsu CY (1993) Expression of c-*fos* and c-*jun* family genes after focal cerebral ischemia. Ann Neurol 33/5:457-464

Angel P, Allegretto EA, Okino ST, Hattori K, Boyle WJ, Hunter T, Karin M (1988) Oncogene *jun* encodes a sequence-specific *trans*-activator similar to AP-1. Nature 332:166-171

Bading H, Ginty, DD, Greenberg ME (1993) Regulation of gene expression in hippocampal neurons by distinct calcium signalling pathways. Science 260:181-186

Ben-Ari Y, Temblay E, Riche D, Ghilini G, Naquet R (1981) Electrographic, clinical and pathological alterations following systemic administration of kainic acid, bicuculline or pentetrazole: metabolic mapping using the deoxyglucose method with special reference to the pathology of epilepsy. Neurosci 6:1361-1391

Berk AJ (1986) Adenovirus promoters and transactivation. Ann Rev Genet 20:45-79

Bohmann D, Bos TJ, Admon A, Nishimura T, Vogt PK, Tjian R (1987) Human proto-oncogene c-*jun* encodes a DNA binding protein with structural and functional properties of transcription factor AP-1. Science 238:1386-1392

Chiasson BJ, Hooper ML, Murphy PR, Robertson HA (1992) Antisense oligonucleotide eliminates *in vivo* expression of c-*fos* in mammalian brain. Europ J Pharm 227:451-453

Cohen D, Curran T (1988) *fra-1*: a serum-inducible, cellular immediate-early gene that encodes a *fos*-related antigen. Mol Cell Biol 8:2063-2069

Colotta F, Polentarutti N, Sironi M, Mantovani A (1992) Expression and involvement of c-*fos* and c-*jun* proto-oncogenes in programmed cell death induced by growth factor deprivation in lymphoid cell lines. J Biol Chem 267/26:18278-18283

Curran T, Morgan JI (1987) Memories of *fos*. Bio Essays 7/6:255-258

Curran T, Teich NM (1982) Candidate product of the FBJ murine osteosarcoma virus oncogene: Characterization of a 55,000 dalton phosphoprotein. J Virology 42:114-122

Dempsey RJ, Carney JM, Kindy MS (1991) Modulation of ornithine decarboxylase mRNA following transient ischemia in the gerbil brain. J Cereb Blood Flow Metab 11:979-985

Dragunow M, Robertson HA (1987) Kindling stimulation induces c-*fos* protein(s) in granule cells of the rat dentate gyrus. Nature 329:441-442

Dragunow M, Robertson HA (1988) Brain injury induces c-*fos* protein(s) in nerve and glial cells in adult mammalian brain. Brain Res 455:295-299

Draisci G, Iadarola MJ (1989) Temporal analysis of increases in c-*fos*, preprodynorphin and preproenkephalin mRNAs in rat spinal cord. Mol Brain Res 6:31-37

Fagarasan MO, Aiello F, Muegge K, Durum S (1990) Interleukin 1 induces ß-endorphin secretion via Fos and Jun in AtT-20 pituitary cells. Proc Natl Acad Sci USA 87:7871-7874

Gentz R, Rauscher FJ III, Abate C, Curran T (1989) Parallel association of Fos and Jun leucine zippers juxtaposes DNA binding domains. Science 243:1695-1699

Goddard GV, McIntyre DC, Leech CK (1969) A permanent change in brain function resulting from daily electrical stimulation. Exp Neurol 25:295-330

Goelet P, Castellucci VF, Schacher S, Kandel ER (1986) The long and the short of a long-term memory - a molecular framework. Nature 322:419-422

Hai T, Curran T (1991) Cross-family dimerization of transcription factors Fos/Jun and ATF/CREB alters DNA binding specificity. Proc Natl Acad Sci USA 88:3720-3724

Hazel TG, Nathans D, Lau LF (1988) A gene inducible by serum growth factors encodes a member of the steroid and thryoid hormone receptor superfamily. Proc Natl Acad Sci USA 85:8444-8448

Hengerer B, Lindholm D, Heumann R, Ruther U, Wagner EF (1990) Lesion-induced increase in nerve growth factor mRNA is mediated by c-*fos*. Proc Natl Acad Sci USA 87:3899-3903

Hershman HR (1989) Extracellular signals, transcriptional responses and cellular specificity. Trends Biochem Sci 14:455-458

Hilberg F, Wagner EF (1992) Embryonic stem (ES) cells lacking functional c-*jun*: consequences for growth and differentiation, AP-1 activity and tumorigenicity. Oncogene 7:2371-2380

Hilbush BS, Curran T, Morgan JI (1994) Cellular immediate-early genes in the nervous system:

genes for all reasons? In: Fuxe K, Agnati LF, Leon A, Ottoson D (eds) Trophic Regulation of the Basal Ganglia. Focus on dopamine Neurons. Pergamon Press, 301-315

Isackson PF, Huntsman MM, Murray KD, Gall CM (1991) BDNF mRNA expression is increased in adult rat forebrain after limbic seizures: temporal patterns of induction distinct from NGF. Neuron 6:937-948

Jacewitz M, Kiessling M, Pulsinelli WA (1986) Selective gene expression of focal cerebral ischemia. J Cereb Blood Flow Metab 6:263-272

Johnson RS, Spiegelman BM, Papaioannou V (1992) Pleiotropic effects of a null mutation in the c-*fos* proto-oncogene. Cell 71:577-586

Jorgenson M, Deckert J, Wright D, Gehlert D (1989) Delayed c-*fos* proto-oncogene expression in the rat hippocampus induced by transient global cerebral ischemia: an *in situ* hybridization study. Brain Res 484:393-398

Kanamatsu T, McGinty JF, Mitchell CL, Hong JS (1986) Dynorphin- and enkephalin-like immunoreactivity is altered in limbic-basal ganglia regions of rat brain after repeated electroconvulsant shock. J Neurosci 6/3:644-649

Kindy M, Verma I (1988) Inhibition of c-*fos* gene expression does not alter the differentiation pattern of PC12 cells. In: Melton D (ed) Antisense DNA and RNA. Cold Spring Harbor Laboratory, ... pp 129-133

Kobierski LA, Chu H-M, Tan Y, Comb MJ (1991) cAMP-dependent regulation of proenkephalin by Jun D and Jun B: Positive and negative effects and AP-1 proteins. Neurobiol 88:10222-10226

Kononen J, Honkaniemi J, Alho H, Koistinaho J, Iadarola M, Pelto-Huikko M (1992) Fos-like immunoreactivity in the rat hypothalamic-pituitary axis after immobilization stress. Endocrin 130/5:3041-3047

Kononen J, Koistinaho J, Alho H (1990) Circadian rhythm in c-*fos*-like immunoreactivity in the rat brain. Neurosci Lett 120:150-108

Konradi, C, Kobierski LA, Nguyen TV, Heckers S, Hyman SE (1993) The cAMP-response-element-binding protein interacts, but Fos protein does not interact, with the proenkephalin enhancer in rat striatum. Proc Natl Acad Sci USA 90:7005-7009

Lamb NJC, Fernandez A, Toukine N, Jeanteur P, Blanchard J-M (1990) Demonstration lin living cells of an intragenic negatie regulatory element within the rodent c-*fos* gene. Cell 61:485-496

Landschulz WH, Johnson PF, McKnight SL (1988) The leucine zipper: a hypothetical structure common to a new class of DNA binding proteins. Science 240:1759-1764

Largent BL, Sosnowski RG, Reed RR (1993) Ionotropic glutamate receptor subtypes activate c-*fos* transcription by distinct calcium-requiring intracellular signalling pathways. Neuron 10:31-41

Lerea LS and McNamara JO (1993) Ionotropic glutamate subtypes activate c-*fos* trancription by distinct calcium-requiring intacellular signalling pathways. Neuron 10:31-41

Lloyd A, Yancheva N, Wasylyk B (1991) Transformation suppressor activity of a Jun transcription factor lacking its activation domain. Nature 352:635-638

Logeat F, Israel N, Ten R, Blank V, Le Bail O, Kourilsky P, Israel A (1991) Inhibition of transcription factors belonging to the rel/NF-κB family by a transdominant negative mutant. EMBO J 10/7:1827-1832

Lucas JJ, Mellstrom B, Colado MI, Naranjo JR (1993) Molecular mechanisms of pain: Serotonin 1A receptor agonists trigger transactivation of c-*fos* of the prodynorphin gene in spinal cord neurons. Neuron 10:599-611

Maki Y, Bos TJ, Davis C, Starbuck M, Vogt PK (1987) Avian sarcoma virus 17 carries the *jun* oncogene. Proc Natl Acad Sci USA 84:2848-2852

Matsui M, Tokuhara M, Konuma Y, Nomura N, Ishizaki R (1990) Isolation of human *fos*-related genes and their expression during monocyte-macrophage differentiation. Oncogene 5:249-255

McNamara JO, Bonhaus DW, Crain BJ, Gellman RL, Shin C (1987) Biochemical and pharmacologic studies of neurotransmitters in the kindling model. In: Jobe PC, Laird HE (eds) Neurotransmitters and Epilepsy. Humana Press, Clifton NJ, pp 115-148

Milbrandt J (1987) A nerve growth factor-induced gene encodes a possible transcriptional regulatory factor. Science 238:797-799

Miller AD, Curran T, Verma IM (1984) c-Fos protein can induce cellular transformation: a novel mechanism of activation of a cellular oncogene. Cell 36:51-60

Molinar-Rode R, Smeyne RJ, Curran T, Morgan JI (1993) Regulation of proto-oncogene expression in adult and developing lungs. Mol Cell Biol 13/6:3213-3220

Morgan JI, Cohen DR, Hemstead JL, Curran T (1987) Mapping patterns of c-*fos* expression in the central nervous system after seizure. Science 237:192-197

Morgan JI, Curran T (1989) Stimulus-transcription coupling in neurons: role of cellular immediate-early genes. TINS 12/11:459-462

Morgan JI, Curran T (1991a) Stimulus-transcription coupling in the nervous system: involvement of the inducible proto-oncogenes *fos* and *jun*. Annu Rev Neurosci 14:421-451

Morgan JI, Curran T (1991b) Proto-oncogene transcription factors and epilepsy. TIPS 12/9:343-349

Murre C, McCaw PS, Vaessin H, Caudy M, Jan LY, Jan YN, Cabrera CV, Buskin JN, Hauschka SD, Lassar AB, Weintraub H, Baltimore D (1989) Interactions between heterologous helix-loop-helix proteins generate complexes that bind specifically to a common DNA sequence. Cell 58:537-544

Nakabeppu Y, Ryder K, Nathans D (1988) DNA binding activities of three murine Jun proteins: stimulation by Fos. Cell 55:907-910

Oberdick J, Smeyne RJ, Jann JR, Zackson S, Morgan JI (1990) A promoter that drives transgenic expression in cerebellar purkinje and retinal bipolar neurons. Science 248:223-226

Okuno H, Suzuki T, Yoshida T, Hashimoto Y, Curran T, Iba H (1991) Inhibition of *jun* transformation by a mutated *fos* gene: design of an antioncogene. Oncogene 6:1491-1497

O'Shea EK, Rutkowski R, Kim PA (1992) Mechanism of specificity in the Fos-Jun oncoprotein heterodimer. Cell 68: 699-708

Patterson SL, Grover LM, Schwarzkroin PA, Bothwell M (1992) Neurotrophin expression in rat hippocampal slices: a stimulus paradigm inducing LTP in CA1 evokes increases in BDNF and NT-3 mRNAs. Neuron 9:1081-1088

Prywes R, Fisch TM, Roeder RG (1988) Transcriptional regulation of c-*fos*. Cold Spr Harb Symp Quant Bio 53:739-748

Rauscher FJ III, Voulalas PJ, Franza BR Jr, Curran T (1988) Fos and Jun bind cooperatively to the AP-1 site: reconstruction in vitro. Genes Dev 2:1687-1699

Riabowol KT, Vosatka RJ, Ziff EB, Lamb NJ, Feramisco JR (1988) Microinjection of *fos*-specific antibodies blocks DNA synthesis in fibroblast cells. Mol and Cell Biol 8:1670-1676

Robertson GS, Herrera DG, Dragunow M, Robertson HA (1989) L-DOPA activates c-*fos* in the striatum ipsilateral to a 6-hydroxydopamine lesion of the substantia nigra. Eur J Pharm 159:99-100

Robertson GS, Robertson HA (1987) D_1 and D_2 dopamine agonist synergism: separate site of action? TIPS 8:295-299

Ruther U, Garber C, Komitowski D, Muller R, Wagner EF (1987) Deregulated c-*fos* expression interferes with normal bone development in transgenic mice. Nature 325:412-418

Ryder K, Lau LF, Nathans D (1988) A gene activated by growth factors is related to the oncogene v-*jun*. Proc Natl Acad Sci USA 85:1487

Ryder K, Lanahan A, Perez-Albuerne E, Nathans D (1989) Jun D: A third member of the Jun gene family. Proc Natl Acad Sci 86:1500-1503

Saffen D, Cole A, Worley P, Christy B, Ryder K, Baraban J (1988) Convulsant-induced increase in transcription factor messenger RNAs in rat brain. Proc Natl Acad Sci USA 85:7795-7799

Sagar SM, Sharp FR, Curran T (1988). Expression of c-*fos* protein in brain: Metabolic mapping at the cellular level. Science 240:1328-1331

Sawai S, Shimono A, Hanaoka K, Kondoh H (1991) Embryonic lethality resulting from disruption of both N-*myc* alleles in mouse zygotes. New Biol 3/9:861-869

Schilling K, Luk D, Morgan JI, Curran T (1991) Regulation of a *fos-lacZ* fusion gene: A paradigm for quantitative analysis of stimulus-transcription coupling. Proc Natl Acad Sci USA 88:5655-5669

Schuermann M, Neuberg M, Hunter JB, Jenuwein T, Ryseck, RP, Bravo R, Muller R (1989) The leucine repeat motif in Fos protein mediates complex formation with Jun/AP-1 and is required for transformation. Cell 56:507-516

Sheng M, McFadden G, Greenberg ME (1990) Membrane depolarization and calcium induce c-*fos* transcription via phosphorylation of transcription factor CREB. Neuron 4:571-582

Shin C, McNamara JO, Morgan JI, Curran T, Cohen DR (1990) Induction of c-*fos* mRNA expression by afterdischarge in the hippocampus or naive and kindled rats. J Neurochem 55/3:1050-1055

Siesjo BK, Bengtsson F (1989) Calcium fluxes, calcium antagonists, and calcium-related pathology in brain ischema, Hypogylcemia, and spreading depression: a unifying hypothesis. J Cereb Blood Flow Metab 9:127-140

Smeyne RJ, Schilling K, Robertson L, Luk D, Oberdick J, Curran T, Morgan JI (1992a) Fos-lacZ transgenic mice: mapping sites of gene induction in the central nervous system. Neuron 8:13 23

Smeyne RJ, Curran T, Morgan JI (1992b) Temporal and spatial expression of a *fos-lac*Z transgene in the developing nervous system. Mol Brain Res 16:158-162

Smeyne RJ, Vendrell J, Hayward M, Baker SJ, Miao GG, Schilling K, Robertson LM, Curran T, Morgan JI (1993) Continuous c-*fos* expression precedes programmed cell death *in vivo*. Nature 363:166-169

Sonnenberg JL, Mitchelmore C, Macgregor-Leon PF, Hempstead J, Morgan JI, Curran T (1989a) Glutamate receptor agonists increase the expression of Fos, Fra and AP-1 DNA binding activity in the mammalian brain. J Neurosci Res 24:72-80

Sonnenberg JL, Rauscher FJ III, Morgan JI, Curran T (1989b) Regulation of proenkephalin by Fos and Jun. Science 246:1622-1625

Sonnenberg JL, Macgregor-Leon PF, Curran T, Morgan JI (1989c) Dynamic alterations occur in the levels and composition of transcription factor AP-1 complexes and seizure. Neuron 3:359-365

Stewart TA, Pattengale PK, Leder P (1984). Spontaneous mammary adenocarinomas in transgenic mice that carry and express MTV/*myc* fusion genes. Cell 38:627-637

Sukhatame VP, Cao X, Chang LC, Tsai-Morris CH, Stamenkovich D, Ferreira PCP, Cohen DR, Edwards SA, Shows TB, Curran T, Le Beau MM, Adamson ED (1988) A zinc finger-encoding gene corregulated with c-*fos* during growth and differentiation, and after cellular depolarization. Cell 53:37-43

Sutula T, Xiao-Xian H, Cavazos J, Scott G (1988) Synaptic reorganization in the hippocampus induced by abnormal functional activity. Science 239:1147-1150

Trugman JM, Wooten GF (1986) The effects of L-DOPA on regional cerebral glucose utilization in rats with unilateral lesions of the substantia nigra. Brain Res. 379:264-274

Turner R, Tijan R (1989) Leucine repeats and an adjacent DNA binding domain mediate the formation of functional cFos-cJun heterodimers. Science 243:1689-1691

Wang ZQ, Ovitt C, Grigoriadis AE, Mohle-Steinlein U, Ruther U, Wagner EF (1992) Bone and haematopioetic defects in mice lacking c-*fos*. Nature 360:741-435

Welsh FA, Moyer DJ, Harris VA (1992) Regional expression of heat shock protein 70 mRNA and c-*fos* mRNA following focal ischemia in rat brain. J Cereb Blood Flow Metab 12:204-212

White JD, Gall CM (1987) Differential regulation of neuropeptide and proto-oncogene mRNA content in the hippocampus following recurrent seizures. Mol Brain Res 3:21-29

Zerial M, Toschi L, Ryseck RP, Schuermann M, Muller R, Bravo R (1989) The product of a novel growth factor activated gene, *fos* B, interacts with JUN proteins enhancing their DNA binding activity. EMBO J 8:805-813

mally inhibits this process. Here we extend these findings and analyze the physiological expression patterns of the two genes in rat brain.

Materials and methods

Tissue preparation

Wistar rats were killed with a lethal dose of Nembutal (90mg/kg i.p.) at postnatal days 5, 14, 40 and 60. Brains were removed from the scull and frozen in cryostat embedding medium (Tissue-Tek). Eighteen-micrometer parasagittal sections were cut on a cryostat and thaw-mounted onto gelatin-coated slides. Sections were then fixed in 4% paraformaldehyde in 1x phosphate-buffered saline (PBS) for 5 min, dipped twice in 1x PBS and acetylated with 0,25% acetic anhydride/0,1% triethanolamine in 0,9% NaCl for 10 min, dehydrated in a graded series of alcohols, delipidated in chloroform for 5 min and stored submerged in 96% ethanol at 4°C.

In situ hybridization

30 mer antisense oligodeoxynucleotide probes, complementary to the c-*jun* and *jun* B genes were labeled to a specific activity of 1 x 10^7 c.p.m./µg using [alpha-^{35}S] dATP (Du Pont). Probe sequences were:

> c-*jun*: TTAGCATGAGTTGGCACCCACTGTTAACGT
> *jun* B: CGTGATAGAAAGCCTGTTCCATTTTCGTGC

The probes were separated from unincorporated nucleotides by spun-column chromatography through Sephadex G-25 (Pharmacia).

Fixated sections were removed from 96% ethanol at 4°C and air-dried. Each section was surrounded by a ring of rubbercement (Marabu) and prehybridized with 50-100 µl of hybridiziation buffer containing 4x sodium chloride-sodium citrate buffer (SSC) (1x SSC = 0.15 M NaCl, 0.015 M sodium citrate, pH 7.0), 50% deionized formamide, 5x Denhardt's solution, 50 mM Na-phosphate, pH 6.5, 1 mM Na-PP, 120 µg/ml heparin, 100 µg/ml sonicated salmon-sperm DNA, 100 µg/ml poly-[A] and 100 µg/ml yeast tRNA for 2 h at room temperature in a humid box. The prehybridization buffer was replaced by 50 µl of hybridization buffer containing 2 x 10^6 c.p.m./100 µl of the labeled probes solubilized in prehybridization buffer. Hybridization was performed overnight at 30°C in a humid box.

Subsequently sections were washed in 1x SSC at room temperature for 30 min and in 0.1 x SSC at 40°C for 1 h with agitation. Slides were dehydrated in a series of graded ethanols containing 300 mM ammonia acetate, air-dried and processed for autoradiography.

Slides were exposed to autoradiographic film (Hyperfilm MP, Amersham) for 30 days at room temperature.

Antisense phosphorothioate oligodeoxynucleotides

Phosphorothioate oligodeoxynucleotides were synthesized on an Applied Biosystems 380B DNA synthesizer using β-cyanoethyl phosphoramidites as previously described and purified by Reverse Phase-HPLC on a C18-column, detritylated with 80% acetic acid, precipitated with 0.3M NaAc/EtOH and dialysed against 0.1M Tris, pH 8.0; sequences were as previously described (Schlingensiepen et al. 1993).

Western blot

Cells were serum starved in RPMI/2% FCS for 3 days, trypsinized and incubated in RPMI/5% FCS/2μM S-ODN for 5 min. 3 x 10^6 cells were plated into 260ml culture flasks and grown for the times indicated in RPMI/5% FCS/2μM S-ODN. Cell extracts were subjected to SDS-polyacrylamide gel electrophoresis, blotted onto Immobilon-P membrane (Millipore) and probed with either a rabbit anti-mouse-c-jun antibody (Oncogene Science, PC07) or a rabbit anti-human-Jun B antibody (Oncogene Science, PC28). Chemiluminescent detection was performed using goat anti-rabbit IgG-alkaline-phosphatase conjugate (Boehringer Mannheim) as second antibody and CSPD for chemiluminescent detection.

Primary rat hippocampal cultures

Hippocampi were removed from the brains of embryonal day 18 rats, dissociated by mild trypsinization and seeded into 24 well plates at a density of 4 x 10^5/ml in a serum free mixture of HM-medium (Honegger) and N2-medium (Boddenstein-Sato) (1:1, v:v), which had been conditioned by primary rat astrocytes and mixed with 50% unconditioned HM/N2-medium. Oligonucleotides were added 1 h after plating. Survival of developing neurons was assayed using the fluorescein-diacetate/ethidium bromide double staining method. Fluorescein-diacetate selectively stains vital cells, while ethidium bromide stains the nuclei of dead cells.

Results

In situ hybridization experiments displayed a virtually complementary expression pattern of the c-*jun* and the *jun* B gene in adult rat brain (Fig. 3). Brain regions which expressed significant levels of *jun* B mRNA like hippocampal region CA1, the cortex and the corpus striatum, displayed little or no c-*jun* expression. In contrast, the dentate gyrus, cerebellum and thalamus contained significant c-*jun* mRNA levels but little *jun* B expression. This highly differential expression pattern in freely moving animals, not stimulated intentionally, indicates that the two homologous genes are activated by different intracellular pathways. Thus interestingly each neuronal region appears to express only one of the two IEGs at significant levels in freely moving animals that were not intentionally stimulated. In order to determine whether the complementary expression pattern is paralleled by significant differences in function between *c-jun* and *jun* B, we analyzed the effects of inhibition of synthesis of the individual proteins.

As described previously, western blot analysis of cell extracts from mouse fibroblasts and mammary carcinoma cells reveals specific suppression of each Jun protein by its respective antisense oligonucleotide, but not by the sequence designed to suppress the other jun family member (Fig. 4). Similar results were obtained in PC12 rat phaeochromocytoma cells (Data not shown). Thus the antisense technique is capable of discriminating between two closely homologous mRNAs of the same gene family and is a highly selective tool for "loss of function analysis" of the respective proteins.

In order to analyze whether the complementary expression patterns of *c-jun* and *jun* B in rat brain correspond with differences in function, we examined the effects of inhibiting Jun B or c-Jun protein expression in differentiating hippocampal neurons.

Postmitotic hippocampal neurons, explanted at embryonal day E18 and transferred into culture were severly altered in their differentiation properties after inhibition of Jun B protein expression (Fig. 5). While after 2 days in culture, no difference was observed between the morphological differentiation patterns of cells treated with oligonucleotides and control conditions, clear alterations in cell morphology were seen after 5 days in culture. Inhibition of Jun B expression resulted in reduced neurite density, while lack of c-Jun expression resulted in a denser neuritic network than control conditions. These differences were even more pronounced after 12 days in culture.

It is also apparent from the morphological picture that concomitant with inhibition of neurite outgrowth the cell numbers appear to be reduced in cultures with

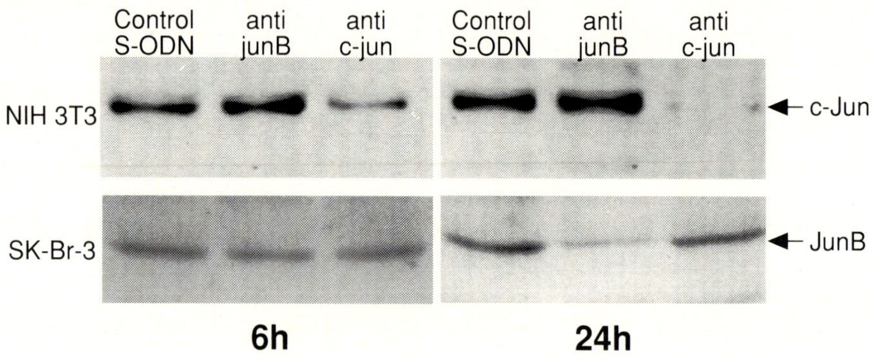

Fig. 4. Selectivity of protein synthesis inhibition with antisense S-ODN. Western blot analysis of cell extracts of NIH 3T3 mouse fibroblasts and of SK-Br-3 human mammary carcinoma cells, treated with either a randomized control oligonucleotide, an anti-c-*jun* S-ODN or an anti-*jun* B-S-ODN. Each of the two specific oligonucleotides inhibits only the respective protein but not the closely homologous other member of the *jun* gene family, demonstrating the specificity of the technique.

reduced Jun B expression. For the purpose of quantification of this effect, we assayed for neuronal survival. Following the inhibition of Jun B expression, neuronal survival was reduced to 66% of controls at day 5 and to 34% at day 12. In contrast, the number of neurons surviving during the process of neuronal development was increased by 44% in anti c-*jun* treated neurons (Fig 6). The fact that in Jun B suppressed neuronal cultures the reduction in cell number and neurite density becomes apparent around day 5 is intriguing since synapse formation starts at this time in hippocampal cultures. For further investigation whether Jun B protein expression may play a role in synapse formation we analyzed *jun* B expression levels during neuronal development. *In situ* hybridization demonstrated little expression of *jun* B mRNA in the cerebral cortex on postnatal day 5 (PND 5) before the beginning of cortical synapse formation. In contrast, a strong peak of expression is seen on PND 14, during the peak phase of synapse formation, declining towards the lower adult levels on PND 40 (Fig. 7).

Discussion

The distribution pattern of c-*jun* and *jun* B expression as shown in our *in situ* hybridization experiments is similar to that described by Mellstöm et al. (1991), however, the expression patterns of the two genes appear even more differential in our experiments.

This may be due to the differences in hybridization probes. The oligonucleotide probes were directed against non-homologous parts of the two genes, while the full length hydrolized cRNA probes used by Mellström et al. must inevitably show some cross-hybridization with homologous parts of the other *jun* family members. Furthermore, the different labelling intensities of the probes and the different exposure times may play a role. Long exposure times obliterate the differences between moderate and high expression levels. Finally, as noted by Mellström et al., some of the expression noted in their study may possibly be due to the interval between removal of animals from their cages and the freezing of brain tissue.

The highly differential and in fact largely complementary expression pattern of the c-*jun* and *jun* B immediate-early genes shows that the two genes must be activated by different pathways. In the adult central nervous system, different neuronal subregions appear to preferentially express only one of the two homologous genes. This is quite different from the situation found in cell culture experiments after serum stimulation and from the effects of massive neuronal stimulation, *e.g.* during seizure induction. The latter experimental conditions obviously activate a great number of intracellular pathways. In contrast, the highly differential expression pattern of the two genes under physiological conditions *in vivo* suggests completely different mechanism of induction. This is consistent with the finding of differential activation by N-methyl-D-aspartate receptor (NMDA-receptor) stimulation in neuronal cell culture, leading to induction of *jun* B, but not of c-*jun* expression (Worley et al. 1990). It is also consistent with the highly differential expression patterns of c-*jun* and *jun* B during organ development (Dungy et

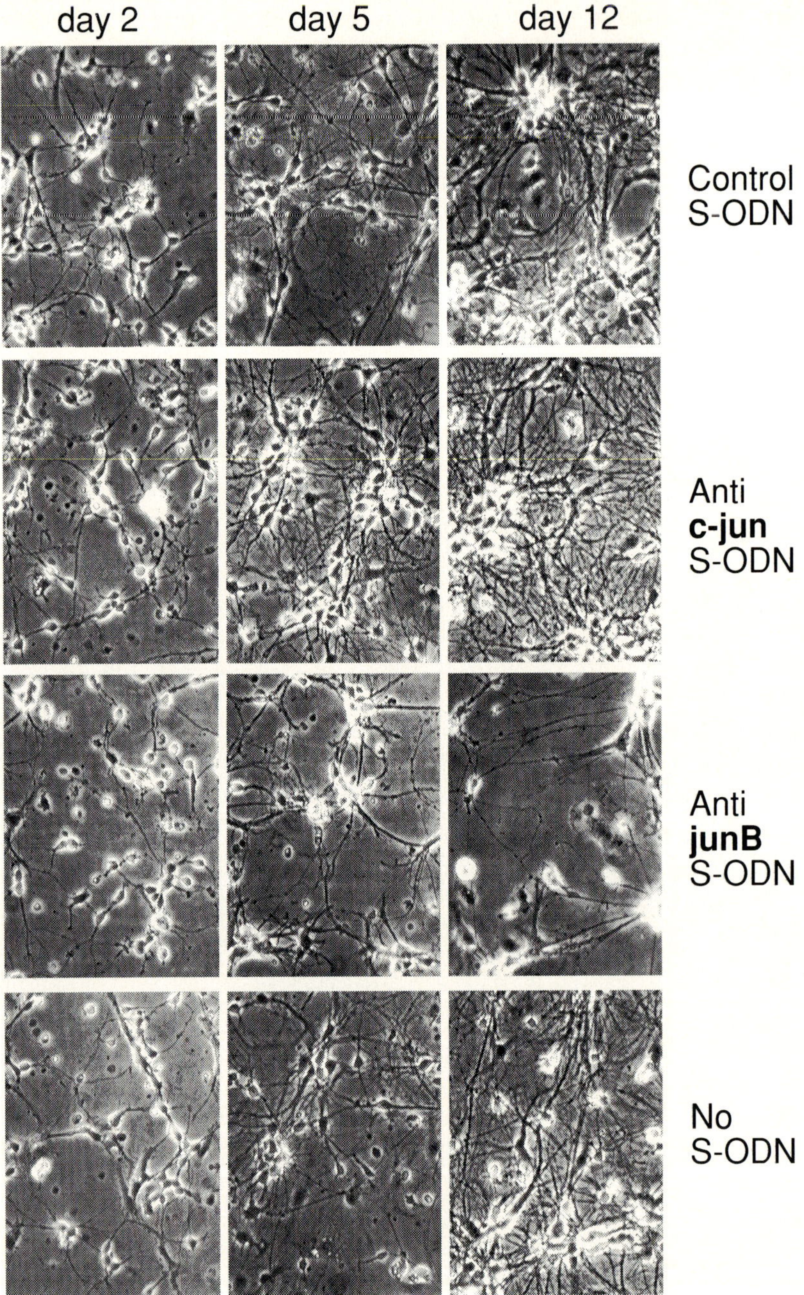

Fig. 5. Suppression of c-Jun or Jun B synthesis in primary neurons. Primary rat hippocampal cell cultures treated either with one of two specific antisense S-ODNs, a randomized control S-ODN or with no oligonucleotide. No significant differences are observed on day 2 of culture. However, 5 days after plating, reduced neurite density is apparent in anti-*jun* B S-ODN treated cells, while the reverse is seen in anti-c-*jun* S-ODN treated neurons. The changes are even more marked on day 12 of hippocampal cell culture.

al. 1991; Wilkinson et al. 1989). Differential expression of different *jun* genes is also seen in nociception experiments and following nerve transection (Herdegen et al. 1991, 1992).

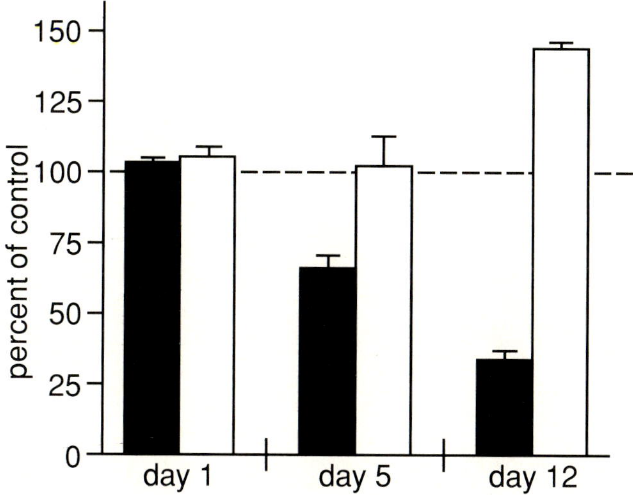

Fig. 6. Survival of developing hippocampal neurons. Neuronal survival of hippocampal neurons after inhibition of c-*jun* or *jun* B expression. White bars: anti-c-*jun* S-ODN; black bars: anti-*jun* B S-ODN. Values are given as percent of controls, treated with randomized control S-ODN. Error bars correspond to 1 SD.

We have previously shown with the antisense technique that c-*jun* and *jun* B have opposite functions in cell proliferation as well as cell differentiation. Suppression of Jun B synthesis increases proliferation and prevents the induction of PC12 cell differentiation, keeping the cells in the proliferative state even in the presence of high levels of NGF (Schlingensiepen et al. 1993). The data presented in this study suggest that the Jun B protein may not only be required for the induction of differentiation, but also during the process of the formation of neural connections. The high expression levels of *jun* B during synapse formation as shown in our *in situ* hybridization experiments in developing brain raise the possibility that the Jun B transcription factor may be required for the induction of genes that are required for synapse formation. Alternatively the beginning of synaptic activity may strongly induce *jun* B expression, which in turn - together with local synaptic events - might play a role in synaptic stabilization. Requirement of Jun B expression for synapse formation and/or stabilization would fit well with our results in primary neuronal culture. In anti-*jun* B-S-ODN treated cells, the reduction in neurite density and neuronal number first becomes apparent after the beginning of synapse formation, which starts around day 4 in hippocampal culture (Seifert et al. 1983). A role in synapse formation and/or stabilization

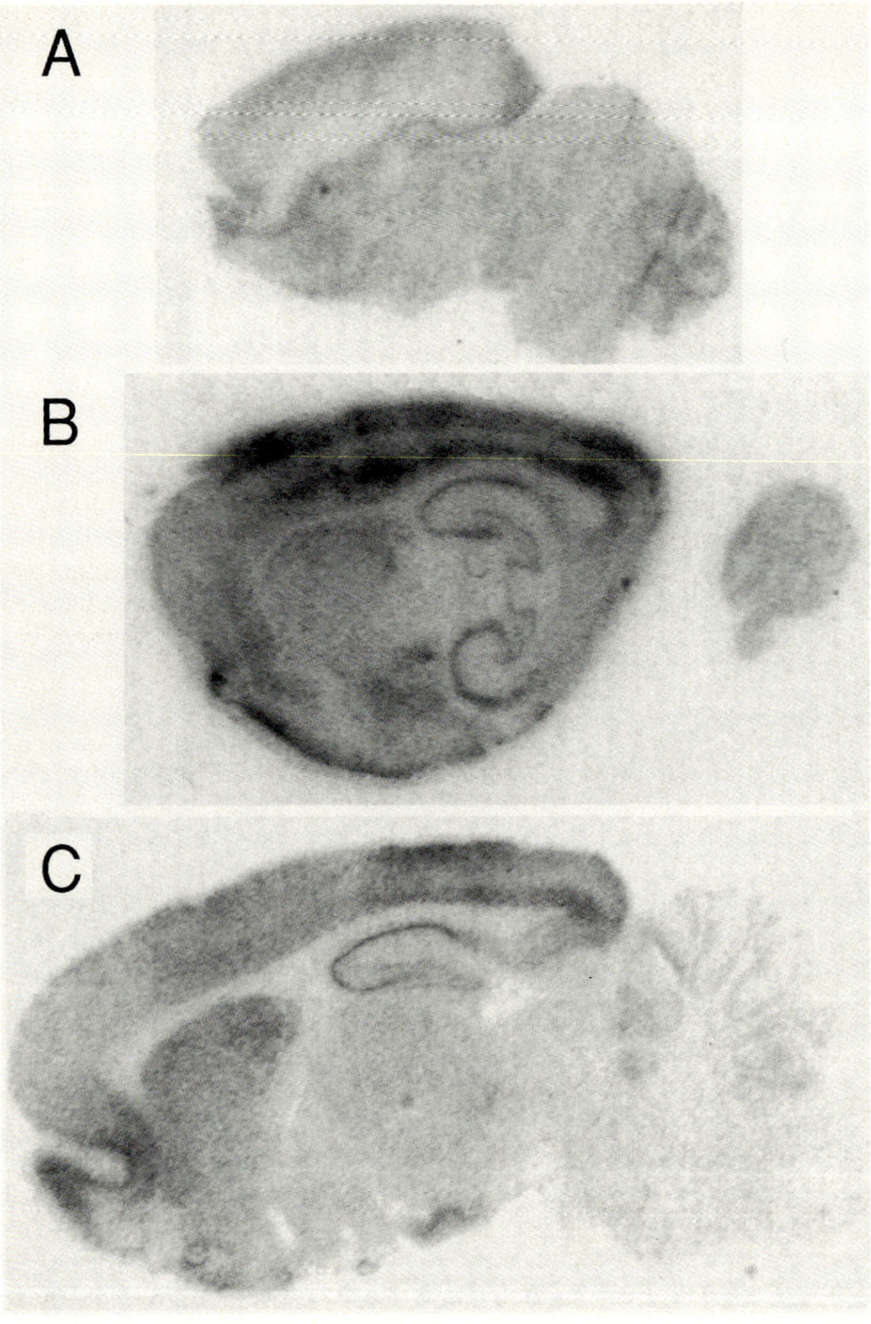

Fig. 7. Expression of *jun* B mRNA in developing rat brain. Autoradiographs from parasagittal sections of rat brain hybridized with antisense oligonucleotide complementary to *jun* B mRNA. **A:** Postnatal day (PND) 5. **B:** PND 14. **C:** PND 40.

also point to a role of Jun B expression in neuronal plasticity. Studies on long term potentiation (LTP), a model for synaptic plasticity, support this concept. After high-frequency stimulation of neurons, leading to LTP induction, expression of *zif/268* and of *jun* B, but not of *c-jun* is induced (Wisden et al. 1990). In contrast, in the same study, commisural stimulation, resulting in inhibition of LTP formation, induced expression of *c-jun* along with the expression of *jun* B. This is consistent with opposite functional roles of the two *jun* genes in neuronal plasticity in analogy to the antagonistic functions in cell proliferation and differentiation (Schlingensiepen et al. 1993). In our hippocampal cell culture experiments, the increase in neuronal survival and neurite density after c-Jun suppression again becomes apparent only after the time point of synapse formation in culture. Since neuronal number and neurite density are increased in cultures treated with anti-c-*jun*-S-ODN, rather than decreased, the balance between the Jun B and c-Jun transcription factors may be involved in mediating signals that regulate competition for neuronal input and for neurotrophic factors, which are required for neuronal survival. Thus the elimination of neurons during certain phases of development, *e.g.* in the primary visual cortex, might involve alterations in the balance of transcription factors such as Jun B and c-Jun. Furthermore they may be involved in regulating the extent of axonal and/or dendritic arborization as indicated by the differences in neurite growth in our cell culture experiments. To further test these hypotheses, *in vivo* studies with the antisense technique will be of great value. Since the balance between the two transcription factors appears to play a crucial role in long term alterations of neuronal structure this leads to the prediction that *in vivo* inhibition of expression of either *jun* gene with antisense molecules should interfere with plasticity, *e.g.* long term memory formation.

Summary

Induction of immediate-early gene encoded transcription factors like Jun B and c-Jun constitutes part of the earliest genetic response to stimuli that induce alterations in cellular programs. We have recently shown that c-Jun and Jun B have opposite functions in cell proliferation as well as in the induction of differentiation. While c-Jun promotes cell growth, Jun B is required for proliferation arrest and the induction of differentiation. Here we show that the expression pattern of the two transcription factors is largely complementary in adult rat brain. Furthermore, using specific inhibition of protein synthesis with antisense phosphorothioate oligodeoxynucleotides we present evidence that *jun* B expression is required for synapse formation and/or stabilization and that c-*jun* functions as an antagonist with respect to this function. Inhibition of *jun* B expression strongly reduces neuronal survival after the time point of initiation of synapse formation, while inhibition of c-*jun* expression increases neuronal survival. These findings suggest that the balance between the two homologous transcription factors is involved in regulating neuronal survival and neuronal cell death and may thus play a role in the elimination of neurons during certain phases of neuronal development and in the accurate generation of neuronal maps.

Acknowledgements:
This work was supported by grants Cr 30/20-1 and Gr 650/5-2 from the Deutsche Forschungsgemeinschaft.

References

Angel P, Allegretto E, Okino S, Hattori K, Boyle W, Hunter T, Karin M (1988) Oncogene jun encodes a sequence-specific transactivator similar to AP-1. Nature 332:166-171

Bartel DP, Sheng M, Lau SF, Greenberg ME (1989) Growth factor and membrane depolarization activate distinct programs of early response gene expression: dissociation of fos and jun induction. Genes and Dev 3:304-313

Behl C, Bogdahn U, Winkler J, Apfel R, Brysch W, Schlingensiepen KH (1993) Autoinduction of platelet derived growth factor-(PDGF) A-chain mRNA and growth inhibition by PDGF-A-chain mRNA-specific phosphorothioate-oligodeoxynucleotides in human malignant melanoma. Biochem Biophys Res Comm 193:744-751

Bohmann D, Bos TJ, Admon A, Nishimura T, Vogt PK, Tjian R (1987) Human proto-oncogene c-jun encodes a DNA binding protein with structural and functional properties of transcription factor AP-1. Science 238:1386-1392

Bravo R (1990) Genes induced during the G0/G1 transition in mouse fibroblasts. Semin Cancer Biol 1:37-46

Chiu R, Angel P, Karin M (1989) Jun B differs in its biological properties from and is a negative regulator of c-Jun. Cell 59:979-986

Dungy LJ, Siddiqi TA, Khan S (1991) C-jun and jun B oncogene expression during placental development. Am J Obstet Gynecol 165:1853-1856

Francastel C, Mazouzi Z, Robert-Lezenes J (1992) Co-induction of c-fos and jun B during the latent period preceding commitment of friend erythroleukemia cells to differentiation. Leukemia 6:935-939

Gerdes W, Brysch W, Schlingensiepen KH, Seifert W (1992) Antisense bFGF oligodeoxynucleotides inhibit DNA synthesis of rat astrocytes. NeuroRep 3:43-46

Herdegen T, Fiallos-Estrada CE, Schmid W, Bravo R, Zimmermann M (1992) The transcription factors c-JUN, JUN D and CREB, but not FOS and KROX-24, are differentially regulated in axotomized neurons following transection of rat sciatic nerve. Mol Brain Res 14:155-165

Herdegen T, Tölle T, Bravo R, Zieglgänsberger W, Zimmermann M (1991) Sequential expression of JUN B, JUN D and FOS B proteins in rat spinal neurons: cascade of transcriptional operations during nociception. Neurosci Lett 129:221-224

Jachimczak P, Bogdahn U, Schneider J, Behl C, Meixensberger J, Apfel R, Dörries R, Schlingensiepen KH, Brysch W (1993) TGF-β_2- specific phosphorothioate-antisense oligodeoxynucleotides may reverse cellular immunosuppression in malignant glioma. J Neurosurg 78:944-951

Kramer IM, Koornneef I, de Vries C, de Groot RP, de Laat SW, van den Eijnden van Raaij AJ, Kruijer W (1991) Phosphorylation of nuclear protein is an early event in TGFβ1 action. Biochem Biophys Res Comm 175:816-822

Laiho M, Ronnstrand L, Heino J, Decaprio JA, Ludlow JW, Livingston DM, Massague J (1991) Control of junB and extracellular matrix protein expression by transforming growth factor-beta 1 is independent of simian virus 40T antigen-sensitive growth-sensitive growth-inhibitory events. Mol Cell Biol 11:972-978

Lamph WW, Wamsley P, Sassone-Corsi P, Verma IM (1988) Induction of proto-oncogene JUN/AP-1 by serum and TPA. Nature 334:629-631

Lau LF, Nathans D (1987) Expression of a set of growth-related immediate-early genes in BALB/c 3T3 cells: Coordinate regulation with c-fos or c-myc. Proc Natl Acad Sci USA 84:1182-1186

Lord KA, Hoffmann-Liebermann B, Liebermann DA (1990) Complexity of the immediate-early response of myeloid cells to the terminal differentiation and growth arrest includes ICAM-1, Jun B and histone variants. Oncogene 5:387-396

Mechta F, Piette J, Hirai SJ, Yaniv M (1989) Stimulation of protein kinase C or protein kinase A mediated signal transduction pathways shows three modes of response among serum inducible genes. New Biol 1:297-304

Mellström B, Achaval M, Montero D, Naranjo JR, Sassone-Corsi P (1991) Differential expression of the jun family members in the rat brain. Oncogene 6:1959-1964

Mollinedo F, Naranjo JR (1991) Uncoupled changes in the expression of the jun family members during myeloid cell differentiation. Eur J Biochem 200:483-486

Morgan JI, Curran T (1991): Human proto-oncogene c-jun encodes a DNA binding protein with structural and functional properties of transcription factor AP-1. Annu Rev Neurosci 14:421-451

Nicolaides NC, Correa I, Casadevall C, Travali S, Soprano KJ, Calabretta B (1992) The Jun family members, c-Jun and Jun D, transactivate the human c-myb promoter via an AP1-like element. J Biol Chem 267:19665-19672

Schütte J, Viallet J, Nau M, Segal S, Fedorko J, Minna J (1989) Jun B inhibits and c-fos stimulates the transforming and transactivating activities of c-jun. Cell 59:987-997

Schlingensiepen KH, Schlingensiepen R, Kunst M, Klinger I, Gerdes W, Seifert W, Brysch W (1993) Opposite functions of Jun B and c-Jun in growth regulation and neuronal differentiation. Dev Genet 14:305-312

Schlingensiepen KH, Brysch W (1992) Phosphorothioate oligomers: Inhibitors of oncogene expression in tumor cells and tools for gene function analysis. In: Erickson R, Izant J (eds) Gene regulation: Biology of Antisense RNA and DNA. Raven, New York, pp 317-328

Seifert W, Rantscht B, Fink HJ, Forster F, Beckh S, Muller HW (1983) Development of hippocampal neurons in cell culture: A molecular approach. In: Seifert W (ed) Neurobiology of the Hippocampus. Academic Press, London, pp 109-135

Sheng M, Greenberg ME (1990) The regulation and function of c-fos and other immediate-early genes in the nervous system. Neuron 4:477-485

Stephens JM, Butts MD, Pekala PH (1992) Regulation of transcription factor mRNA accumulation during 3T3-L1 preadipocyte differentiation by tumor necrosis factor-alpha. J Mol Endocrinol 9:61-72

Vogt PK, Bos TJ (1990) Jun: Oncogene and transcription factor. Adv Cancer Res 55:1-35

Wilkinson DG, Bhatt S, Ryseck RP, Bravo R (1989) Tissue-specific expression of c-jun and jun B during organogenesis in the mouse. Development 106:465-471

Wisden W, Errington ML, Williams S, Dunnett SB, Waters C, Hitchcock D, Evan G, Bliss TVP, Hunt SP (1990) Differential expression of immediate-early genes in the hippocampus and spinal cord. Neuron 4:603-614

Worley PF, Cole AJ, Murphy TH, Christy BA, Nakabeppu Y, Baraban JM (1990) Synaptic regulation of immediate-early genes in brain. Cold Spring Harbor Symposia on Quantitative Biology LV: 213-223

Wu BY, Fodor EJ, Edwards RH, Rutter WJ (1989) Nerve growth factor induces the proto-oncogene c-jun in PC12 Cells. J Biol Chem 264:9000-9003

Characterization and expression of *rel*B, a new member of the *rel*/NF-κB family of transcription factors

R.-P. Ryseck, D. Carrasco, and R. Bravo
Department of Molecular Biology, Bristol-Myers Squibb Pharmaceutical Research Institute. PO Box 4000, Princeton, New Jersey 08543-4000

Introduction

It is well established that induction of cell proliferation by growth factors and other mitogens requires the synthesis of new mRNA. During the last years several laboratories have searched for the so called immediate-early or early response genes which are transcriptionally activated during the G_0 to G_1 transition, independently of *de novo* protein synthesis (for review see Bravo 1990; Herschman 1991; Lau and Nathans 1991). The identification of growth factor-responsive genes is essential for the understanding of cell proliferation and other normal processes governed by growth factors such as differentiation and wound healing.

More than one hundred immediate-early genes have been isolated, but the function for many of them is still unknown (for review see Bravo 1990; Herschman 1991; Lau and Nathans 1991). Among those so far identified are some cytokines, phosphatases, kinases, cytoskeletal matrix proteins, and many different transcription factors. The last group includes members of the Fos and Jun families, proteins containing the helix-loop-helix or the zinc finger motifs, and others containing less defined motifs like the Rel homology domain (RHD) found in the proto-oncogene c-*rel* protein product and the transcription factor complex NF-κB.

The proto-oncogene c-*rel* (Wilhelmsen et al. 1984) is the cellular homolog of the v-*rel* oncogene isolated from reticuloendotheliosis virus strain T, a turkey derived acutely transforming retrovirus causing lymphoid leukemia (Theilen et al. 1966). These were the first identified members of a recently discovered gene family that share a conserved region of approximately 300 amino acids (RHD), which includes sequences important for dimerization, DNA binding, and nuclear localization (for reviews see Gilmore 1990, 1991, 1992; Baeuerle 1991; Baeuerle and Baltimore 1991; Blank et al. 1992; Nolan and Baltimore 1992; Grilli et al. 1993). Another early member of this family is the maternal gene *dorsal* of *Drosophila melanogaster* involved in establishing the dorsal-ventral polarity in the developing embryo (Steward 1987). The more recently described members of this family include NFKB1 (p105/p50-NF-κB, also known as KBF1; Bours et al. 1990; Ghosh et al. 1990; Kieran et al. 1990; Meyer et al. 1991), NFKB2 (p100/p52-NF-κB, also known as p50B, p49 and *lyt*-10; Neri et al. 1991; Schmid et al. 1991; Bours et al. 1992), *rel*A (p65-NF-κB; Nolan et al. 1991; Ruben et al. 1991) and *rel*B (Ryseck et al. 1992). The original NF-κB activity was identified as being capable to bind to the κB site in the immunoglobulin κB light chain enhancer (for review see Lenardo and Baltimore 1989). Although initially the NF-κB activity was thought to consist only of the p65- and p50-NF-κB subunits, it was subsequently shown that all the Rel proteins have the capacity to form various homo- and heterodimers able to bind to κB sites present in the regulatory regions of many genes.

Another characteristic shared by the members of the Rel family is their regulated translocation from the cytoplasm to the nucleus regulated by their interaction with members of the IκB family (for review see Nolan and Baltimore 1992). After stimulation of cells with different agents like lipopolysaccharide or serum, NF-κB complexes dissociate from IκB and translocate to the nucleus where they function as transcriptional activators. Recent *in vivo* studies have demonstrated that IκBα can indeed regulate the translocation of p50/p65 heterodimers and p65 homodimers to the nucleus, but not that of p50 homodimers (Zabel et al. 1993). *In vitro*, IκB has the effect of preventing NF-κB from binding to DNA.

Here we will report some recent studies mainly on one of the members of the Rel family, RelB.

Materials and Methods

Cloning and sequencing of *rel*B

The isolation and characterization of the *rel*B clone was described before (Ryseck et al. 1992).

Generation of antibodies

Different regions of the Rel and IκB proteins were cloned in the pEx34 vector (Strebel el at. 1988) to generate MS2 polymerase fusion protein which after purification was used as antigens in rabbits to generate specific antibodies. Sera samples were tested for their ability to immunoprecipitate *in vitro* translated proteins. The antibodies were checked by immunoprecipitation against the other *in vitro* translated proteins belonging to the same family for cross-reactivity.

Cell culture, labeling, and immunoprecipitation

A detailed description of these procedures appears in Kovary and Bravo 1991.

Transfections

Transfections and chloramphenicol acetyltransferase (CAT) assays were performed as described (Dobrzanski et al. 1993).

In situ hybridization. Embryos were obtained from natural matings of B6C3F1 mice. After dissection, the embryos or tissues were transferred to O.C.T. embedding medium (Miles laboratories) and then mounted on a cryostat specimen holder. The embedded samples were frozen by immersing in liquid nitrogen. Samples were kept in the cryostat at -18°C for at least 1 hour to equilibrate to the appropriate temperature before sectioning. Cryostat sections of 8 to 10 μm thickness were collected on subbed slides (Fisher Scientific Co.), air dried at room temperature for 30 minutes and then stored at -70°C prior to use. Specific cDNA probes were labeled with [^{35}S]UTP from linearized template DNAs, using T3 or T7 RNA polymerases (Promega). Probes were degraded to an average size of 200 nucleotides by limited alkaline hydrolysis. After removal of unincorporated nucleotide using a Sephadex G-50 column (Pharmacia), the labeled probes were stored at -80°C in a solution containing 10 mM DTT and 1 mg/ml yeast tRNA.

In vitro hybridization was performed essentially as previously described by Dony and Gruss (1987). After developing, the samples were dehydrated, mounted with coverslips and microphotographed under dark-field. For bright-field photography, sections were stained with hematoxylin.

Results

Homology of RelB with other family members

Among our collection of cDNA clones corresponding to growth factor inducible genes (Almendral et al. 1988), one was identified because of its striking similarity to the c-*rel* proto-oncogene. Subsequently a full length cDNA clone was isolated which encoded a predicted protein of 558 amino acids containing a region of approximately 300 amino acids (aa 103 to 391), which presents a high identity to v-Rel, c-Rel, p50-NF-κB, p65-NF-κB, p50B-NF-κB, and the *Drosophila* maternal effect protein Dorsal (Fig. 1). The highest identity of 52% in this region was found with c-Rel, therefore, the gene was named *rel*B. At the end of this homology region, known as the Rel homology domain (RHD), RelB contains a positively charged sequence that is similar to the nuclear localization sequence identified in v-Rel and present in other Rel family proteins. A putative site for serine phosphorylation by protein kinase A (RRXS), present in the other members of the family with the exception of p50B, is also absent in RelB having instead the sequence QRLT (aa 360 to 363).

Expression of RelB

The expression of c-*rel* was initially demonstrated to be transiently induced by stimulation of quiescent NIH3T3 cells and to be superinduced in the presence of cycloheximide (Bull et al. 1989).

To study the expression of RelB and to compare it with other members of the family, we generated specific antisera against fusion proteins expressed in bacteria using unique areas of the different Rel family members and IκBα. These antisera were used to determine the *in vivo* expression of RelB by immunoprecipitation under denaturing conditions of [^{35}S]methionine labeled quiescent or serum stimu-

149

lated NIH3T3 cells followed by gel electrophoresis. We were able to show a
significant increase in synthesis following serum stimulation of resting NIH3T3
cells (Ryseck et al. 1992).

```
RelB     PYLVITEQPKQRGMRFRYECEGRSAGSILGESSTEASKTQPAIELRDCGGLREVEV  158
C-Rel    PYVEIIEQPRQRGMRFRYKCEGRSAGSIPGERSTDNNRTYPSVQIMNYYGKGKIRI   63
p65      PYVEIIEQPKQRGMRFRYKCEGRSAGSIPGERSTDTTKTHPTIKINGYTGPGTVRI   74
p50      PYLQILEQPKQRGFRFRYVCEGPSHGGLPGASSEKNKKSYPQVKICNYVGPAKVIV   95
p50B     PYLVIVEQPKQRGFRFRYGCEGPSHGGLPGASSEKGRKTYPTVKICNYEGPAKIEV   93
Dorsal   PYVKITEQPAGKALRFRYECEGRSAGSIPGVNSTPENKTYPTIEIVGYKGRAVVVV  102


RelB     TACLVWKDWPHRVHPHSLVGKD.CTD.GVCRVRLRPHVSPRHSFNNLGIQCVRKKE  212
C-Rel    T..LVTKNDPYKPHPHDLVGKD.CRD.GYYEAEFGPERRP.LFFQNLGIRCVKKKE  214
p65      S..LVTKDPPHRPHPHELVGKD.CRD.GYYEADLCPDRSI.HSFQNLGIQCVKKRD  125
p50      Q..LVTNGKNIHLHAHSLVGKH.CED.GVCTVTAGPKDMV.VGFANLGILHVTKKK  146
p50B     D..LVTHSDPPRAHAHSLVGKQ.CSELGICAVSVGPKDMT.AQFNNLGVLHVTKEN  145
Dorsal   S..CVTKDTPYRPHPHNLVGKEGCKK.GVCTLEINSETMR.AVFSNLGIQCVKKKD  154


RelB     IEAAIERKIQLG.IDPYNAG..........................SLKNH  236
C-Rel    VKEAIILRISAG.INPFNVG..........................EQQLLDI  140
p65      LEQAISQRIQTN.NNPFHVP..........................IEEQR  149
p50      VFETLEARMTEACIRGYNPGLLVHSDLAYLQAEGGGDRQLTDREKEIIRQAAVQQT  202
p50B     MMGTMIQKLQRQRLRSRPQG...................LTEAEQRELEQEAKELK  183
Dorsal   IEAALKAR.EEIRVDPFKTG..........................FSHRFQP  180


RelB     QEVDMNVVRICFQASY.RDQQGHL.HRMDPILSEPVYDKKSTNTSELRICRINKES  290
C-Rel    EDCDLNVVRLCFQVFL.PDEHGNFTTALPPIVSNPIYDNRAPNTAELRICRVNKNC  195
p65      GDYDLNAVRLCFQVTV.RDPAGRP.LLLTPVLSHPIFDNRAPNTAELKICRVNRNS  203
p50      KEMDLSVVRLMFTAFL.PDSTGSFTRRLEPVVSDAIYDSKAPNASNLKIVRMDRTA  257
p50B     KVMDLSIVRLRFSAFL.RASDGSFSLPLKPVISQPIHDSKSPGASNLKISRMDKTA  239
Dorsal   SSIDLNSVRLCFQVFMESEQKGRFTSPLPPVVSEPIFDKKA..MSDLVICRLCSCS  234


RelB     GPCTGGEELYLLCDKVQKEDISVVF......STASWEGRADFSQADVHRQIAIVFKT  341
C-Rel    GSVRGGDEIFLLCDKVQKDDIEVRF......VLNDWEARGVFSQADVGRQVAIVFKT  246
p65      GSCLGGDEIFLLCDKVQKEDIEVYF......TGPGWEARGSFSQADVHRQVAIVFRT  289
p50      GCVTGGEEIYLLCDKVQKDDIQIRFYEEEENGGVWEGFGDFSPTDVHRQFAIVFKT  313
p50B     GSVRGGDEVYLLCDKVQKDDIEVRFYEDDENG..WQAFGDFSPTDVHKQYAIVFRT  292
Dorsal   ATVFGNTQIILLCEKVAKEDISVRFFEEKNGQSVWEAFGDFQHTDVHKQTAITFKT  290


RelB     PPYEDLEISEPVTVNVFLQRLTDGVCSEPLPFTYLPRDHDSYGVDKKRKR  391
C-Rel    PPYCK.AILEPVTVKMQLRRPSDQEVSESMDFRYLPDEKDAYANKSKKQK  295
p65      PPYADPSLQAPVRVSMQLRRPSDRELSEPMEFQYLPDTDDRHRIEEKRKH  304
p50      PKYKDVNITKPASVFVQLRRKSDLETSEPKPFLYYPEIKDKEEVQRKRQK  363
p50B     PPYHKMKIERPVTVFLQLKRKRGGDVSDSKQFTYYPLVEDKEEVQRKRRK  341
Dorsal   PRYHTLDITEPAKVFIQLRRPSDGVTSEALPFEYVPMDSDPAHLRRKRQK  340
```

Fig. 1. Comparison of RelB with other members of the Rel family. Only the region with high similarity among the Rel family is shown. Stippled boxes represent identity among all six proteins, and
stippling alone represents identity among four of the six proteins compared. Mouse sequences: c-Rel
(Bull et al. 1990), p65-NF-κB (Nolan et al. 1991) and p50-NF-κB (Ghosh et al. 1990). Human sequence: p50B-NF-κB (Bours et al. 1992). *Drosophila* Dorsal was taken from Steward, 1987.

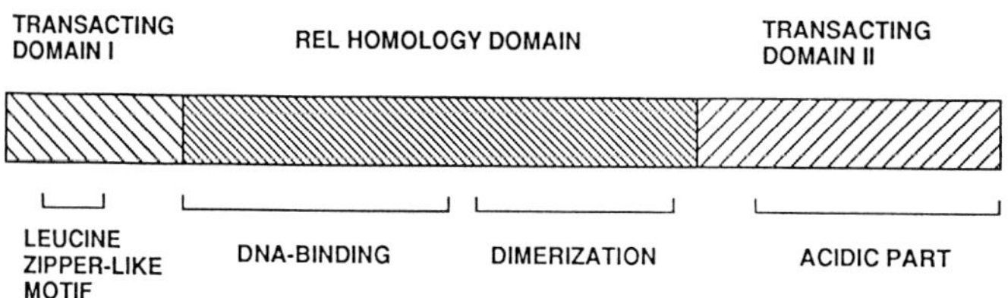

Fig. 5. Summary of the different functional domains of RelB. The N- and C-terminal regions contain the two transactivation domains. The extent of the DNA binding and the dimerization domains as determined by deletion analysis in the RHD is indicated.

Specific expression of relB transcripts during lymphoid organ development

With the exception of c-*rel*, little is known about the tissue specific expression of the *rel* family genes in vertebrates. The mammalian c-*rel* transcript is expressed at highest levels in B and T lymphoid cells in mouse and human. The cytoplasmic NF-κB DNA binding activity has been found in many tissues and cell types suggesting a widespread expression of p65- and p50-NF-κB, although the composition of the complexes containing NF-κB activity has still to be determined.

The different DNA binding and transcriptional activities of Rel proteins suggest that they could play differential roles during development with a selective expression in particular tissues. In an attempt to address this question and to better understand the biological role of RelB, we studied the expression of the *rel*B gene during mouse development by *in situ* hybridization and compared it to the other members of the *rel* family and IκBα.

In situ hybridization in sagital sections of day 14 embryos revealed that transcripts for p50- and p65-NF-κB, and IκBα are present at significant levels in nearly all tissues, the highest level of expression for these genes being in the thymus (data not shown). On the other hand, p50B-NF-κB, c-*rel*, and *rel*B transcripts are present at very low or undetectable levels in all tissues with the exception of the thymus, where a very weak signal can be observed. A similar pattern of expression for these genes is observed in sagital sections of 17 day old embryos. Observation at higher magnification of the embryonic thymus at days 14 and 17 reveals that the pattern of expression of *rel*B and c-*rel* changes as the tissue develops. In the thymus of 14 day old embryos, the *rel*B and c-*rel* transcripts are homogeneously distributed throughout the tissue (not shown). In contrast to this, the hybridization signals are heterogeneously distributed in the thymus of 17 day embryos. The transcripts arc highly expressed in specific areas of the tissue while in other regions there is no expression at all. These differences in expression correlate with the changes in the cellular composition of the thymus during development. By day 14 of fetal life, the thymus is a simple lobular structure composed mainly of epithelial cells and some thymocyte precursors derived from the fetal liver. After day 17, the thymus begins to show cortical and medullar differentiation. At this time p50- and p65-NF-κB, and IκBα mRNAs are strongly expressed in thymus with a preference for cortical structures. The expression of *rel*B and c-*rel* is very low but detectable in the medullar regions (data not shown).

The observation that *rel*B transcripts are detected at significant levels in adult thymus but are weakly observed in the thymus of day 17 embryos indicates that their expression possibly increases after birth. Results obtained by *in situ* hybridization of sagittal sections of the thymus from new born and older animals with a *rel*B specific probe confirmed that the expression of *rel*B rapidly increases after birth (not shown) and that it is significantly expressed in adult thymus (Fig. 6).

To identify the region of the adult thymus where *rel*B, p50-, p65-NF-κB, and IκBα are expressed, the corresponding sections used for *in situ* hybridization were later stained with hematoxylin (Fig. 6). After staining, two regions can be clearly

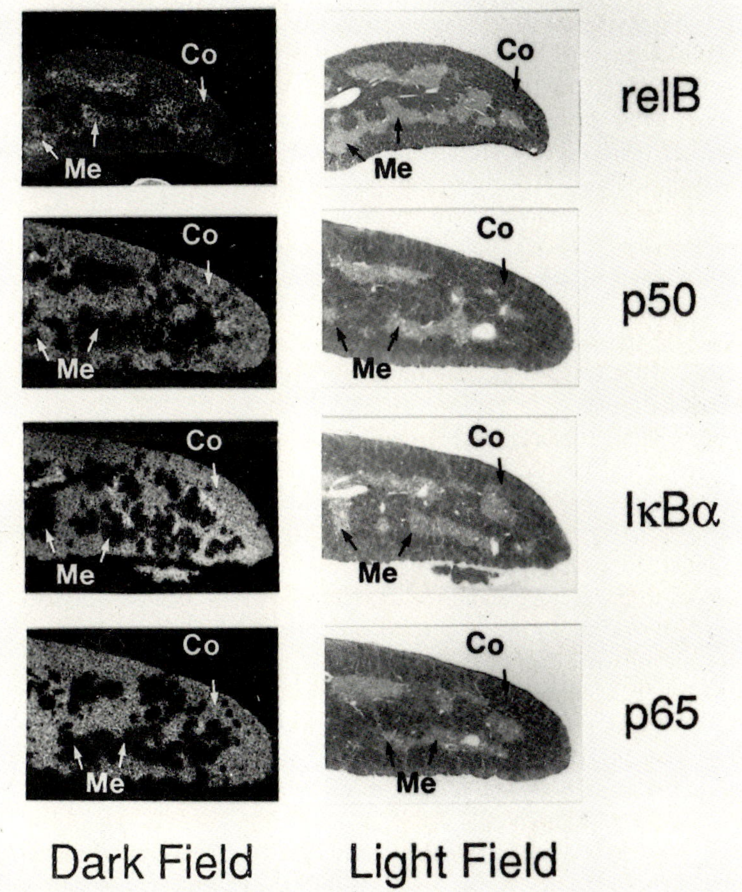

Fig. 6. *In situ* localization of *rel*B, p50-NF-κB, p65-NF-κB, and IκBα mRNAs in adult mouse thymus. Tissue sections from six weeks old mice were hybridized with specific antisense [^{35}S]riboprobes and photographed under dark-field illumination (left panels). After hematoxylin staining, sections were photographed under light-field illumination (right panels). Me = Medulla; Co = Cortex.

distinguished. The cortex, as a dark staining region tightly packed with small thymocytes, and the medulla, as a paler staining central region more loosely packed with slightly larger thymocytes. Interestingly, the results demonstrate that *rel*B is expressed in the medulla, while no signals are detectable in the cortex. In contrast, p50-, p65-NF-κB, and IκBα mRNAs are found mainly in the cortical areas, demonstrating that indeed the different members of the *rel* family are expressed in different regions of the thymus, and probably in various cell types.

Discussion

We have identified a new *rel*-related gene, *rel*B, which belongs to the set of immediate-early genes in fibroblasts. The extent of similarity in the Rel homology domain between RelB and the other members of the family suggests that these proteins probably have similar structures in this region. RelB, as other members of the family is a very stable molecule. When complexed with p50- or p50B-NF-κB, RelB is able to bind with high affinity to different κB sites as determined by electrophoretic mobility shift assays. Most importantly, the complexes RelB/p50-NF-κB and RelB/p50B-NF-κB strongly transactivate *in vivo* a promoter driven by a κB site, similar to complexes containing p65-NF-κB.

RelB has a number of unique characteristics with respect to the other Rel proteins, specially to p65-NF-κB and c-Rel, its closest homologs. For instance, in RelB, the N-terminal region preceding the Rel homology domain is much longer and contains a putative leucine zipper-like motif. Our experiments show that alterations in the structure of this motif significantly decrease the transcriptional capacity of RelB. Furthermore, the C-terminus of RelB is not sufficient to confer full transcriptional activity, in contrast to c-Rel (Bull et al. 1990; Kamens et al. 1990) and p65-NF-κB (Schmitz and Baeuerle 1991; Ruben et al. 1992) whose C-termini are fully responsible for their transcriptional activation potential. This suggests that different regions of RelB must simultaneously interact with several proteins involved in transcriptional activation to exert its function. The identification of the putative factors interacting with the N- and C-terminal regions of RelB will be essential to better understand the mechanisms controlling its activity.

Another feature specific for RelB is its low capacity to form homodimers and its restricted interaction with the other members of the family. RelB only efficiently associates with p50- and p50B-NF-κB. In contrast, all the other members of this family are able to form homo- and heterodimers with each other. Although the Rel homology domain is quite similar, our results suggest that the region of RelB involved in dimerization has some unique characteristics. Further studies are needed to understand these differences.

Our finding that the expression of the *rel*B mRNA is significantly different to the other members of the *rel* family further confirms the unique character of this gene. The expression pattern of *rel*B suggests that its protein product plays a specific role in the development and/or maturation of hematopoietic cells. The highest levels of *rel*B mRNA in the adult mice are detected in the spleen and thymus. While p50- and p65-NF-κB transcripts are detected very early during the embryonic development and are ubiquitously expressed, *rel*B transcripts are only found in the embryonic thymus at late stages of development. Its level of expression increases slowly reaching adult levels 6 days after birth. Interestingly, the expression of *rel*B is confined to the thymic medulla, while p50- and p65-NF-κB, and their inhibitor IκBα are mainly expressed in the cortical structures of the thymus. The identification of the specific cell type expressing RelB and its role in the development of these cells are currently under investigation.

Although a significant progress has been made in the characterization of RelB, its physiological role remains unknown. An important question to be answered, after identifying the cells expressing RelB, is the identification of specific target genes. It will be interesting to analyze how different members of the Rel family are able to discriminate their different target genes. As important will be to investigate the mechanisms governing the expression and activity of the *rel*B gene and its protein product.

Summary

We have identified a serum-inducible gene, *rel*B, which encodes a protein of 558 amino acids containing a region with high similarity to c-Rel and other members of the Rel family. RelB can stimulate promoter activity in the presence of p50-NF-κB in mammalian cells. Transcriptional activation studies demonstrate that the N- and C-termini of RelB are required for full transactivation in the presence of p50-NF-κB. Furthermore, mutational analysis shows that the integrity of the leucine zipper-like motif present in the N-terminus of RelB is essential for its transcriptional activity. The region of RelB required for its interaction with p50-NF-κB has been mapped to the last 110 amino acids of its Rel homology domain. The expression of the *rel*B gene during mouse development is highly restricted to a subpopulation of cells that colonize the lymphoid tissues. *rel*B transcripts are weakly detected in the thymus at late stages of embryogenesis but their level of expression significantly increases after birth. While *rel*B is found in the medullary region of the thymus, p50-, p65-NF-κB, and IκBα are mainly expressed in the cortex.

References

Almendral JM, Sommer D, Macdonald-Bravo H, Burckhardt J, Perera J, Bravo R (1988) Complexity of the early genetic response to growth factors in mouse fibroblasts. Mol Cell Biol 8:2140-2148

Baeuerle PA (1991) The inducible transcription activator NF-κB: regulation by distinct protein subunits. Biochim Biophys Acta 1072:63-80

Baeuerle PA, Baltimore D (1991) The physiology of the NF-κB transcription factor. In: Cohen P, Foulkes JG (eds) The hormonal control regulation of gene transcription. Vol 6, Elsevier, Amsterdam, pp 409-432

Blank V, Kourilsky P, Israel A (1992) NF-κB and related proteins: Rel/dorsal homologies meet ankyrin-like repeats. Trends Biochem Sci 17:135-140

Bours V, Burd PR, Brown K, Villalobos J, Park S, Ryseck R-P, Bravo R, Kelly K, Siebenlist U (1992) A novel mitogen-inducible gene product related to p50/p105-NF-κB participates in transactivation through a κB site. Mol Cell Biol 12:343-350

Bours V, Villalobos J, Burd PR, Kelly K, Siebenlist U (1990) Cloning of a mitogen-inducible gene encoding a κB DNA-binding protein with homology to the *rel* oncogene and to cell-cycle motifs. Nature 348:76-80

Bravo R (1990) Growth factor-responsive genes in fibroblasts. Cell Growth Differ 1:305-309

Bull P, Hunter T, Verma I (1989) Transcriptional induction of the murine c-*rel* gene with serum and phorbol-12-myristate-13-acetate in fibroblasts. Mol Cell Biol 9:5239-5243

Bull P, Morley KL, Hoekstra MF, Hunter T, Verma IM (1990) The mouse c-*rel* protein has an N-terminal regulatory domain and a C-terminal transcriptional transactivation domain. Mol Cell Biol 10:5473-5485

Dobrzanski P, Ryseck R-P, Bravo R (1993) Both N- and C-terminal domains of RelB are required for full transcativation: Role of the N-terminal leucine zipper-like motif. Mol Cell Biol 13:1572-1582

Dony C, Gruss P (1987) Proto-oncogene c-*fos* expression in growth regions of fetal bone and mesodermal web tissue. Nature 328:711-714

Ghosh S, Gifford AM, Riviere LR, Tempst P, Nolan GP, Baltimore D (1990) Cloning of the p50 DNA binding subunit of NF-κB: homology to *rel* and *dorsal*. Cell 62:1019-1029

Gilmore TD (1990) NF-κB, KBF1, *dorsal*, and related matters. Cell 62:841-843

Gilmore TD (1991) Malignant transformation by mutant Rel proteins. Trends Genet 7:318-322

Gilmore TD (1992) Role of *rel* family genes in normal and malignant lymphoid cell growth. Cancer Surveys 15:69-87

Grilli M, Chiu J-S, Lenardo MJ (1993) NF-κB and Rel-participants in a multiform transcriptional regulatory system. Int Rev Cytol 148:1-63

Herschman HR (1991) Primary response genes induced by growth factors and tumor promoters. Annu Rev Biochem 60:281-319

Kamens J, Richardson P, Mosialos G, Brent R, Gilmore T (1990) Oncogenic transformation by v-*rel* requires an amino-terminal activation domain. Mol Cell Biol 10:2840-2847

Kieran M, Blank V, Logeat F, Vandekerckhove J, Lottspeich F, LeBail O, Urban MB, Kourilsky P, Baeuerle PA, Israel A (1990) The DNA binding subunit of NF-κB is identical to factor KBF1 and homologous to the *rel* oncogene product. Cell 62:1007-1018

Kovary K, Bravo R (1991) Expression of different Jun and Fos proteins during the G0 to G1 transition in mouse fibroblasts: *In vitro* and *in vivo* associations. Mol Cell Biol 11:2451-2459

Lau LF, Nathans D (1991) Genes induced by serum growth factors. In: Cohen P, Foulkes JG (eds) The hormonal control regulation of gene transcription. Vol 6, Elsevier, Amsterdam pp 165-201

Lenardo MJ, Baltimore D (1989) NF-κB: a pleiotropic mediator of inducible and tissue-specific gene control. Cell 58:227-229

Meyer R, Hatada EN, Hohmann HP, Haiker M, Bartsch C, Rothlisberger U, Lahm HW, Schlaeger EJ, Van Loon AP, Scheidereit C (1991) Cloning of the DNA-binding subunit of human nuclear factor κB: the level of its mRNA is strongly regulated by phorbol ester or tumor necrosis factor alpha. Proc Natl Acad Sci USA 88:966-970

Neri A, Chang CC, Lombardi L, Salina M, Corradini P, Maiolo AT, Chaganti RS, Dalla FR (1991) B cell lymphoma-associated chromosomal translocation involves candidate oncogene *lyt*-10, homologous to NF-κB p50. Cell 67:1075-1087

Nolan GP, Baltimore D (1992) The inhibitory ankyrin and activator Rel proteins. Curr Opin Genet Develop 2:211-220

Nolan GP, Ghosh S, Liou HC, Tempst P, Baltimore D (1991) DNA binding and IκB inhibition of the cloned p65 subunit of NF-κB, a *rel*-related polypeptide. Cell 64:961-969

Ruben SM, Dillon PJ, Schreck R, Henkel T, Chen CH, Maher M, Baeuerle PA, Rosen CA (1991) Isolation of a *rel*-related human cDNA that potentially encodes the 65-kD subunit of NF-κB. Science 251:1490-1493

Ruben SM, Narayanan R, Klement JF, Chen C-H, Rosen CA (1992) Functional characterization of the NF-κB p65 transcriptional activator and an alternatively spliced derivative. Mol Cell Biol 12:444-454

Ryseck RP, Bull P, Takamiya M, Bours V, Siebenlist U, Dobrzanski P, Bravo R (1992) RelB, a new Rel family transcription activator that can interact with p50-NF-κB. Mol Cell Biol 12:674-684

Schmid RM, Perkins ND, Duckett CS, Andrews PC, Nabel GJ (1991) Cloning of an NF-κB subunit which stimulates HIV transcription in synergy with p65. Nature 352:733-736

Schmitz ML, Baeuerle PA (1991) The p65 subunit is responsible for the strong transcription activating potential of NF-κB. EMBO J 10:3805-3817

Steward R (1987) *Dorsal*, an embryonic polarity gene in Drosophila, is homologous to the vertebrate proto-oncogene, c-*rel*. Science 238:692-694

Theilen GH, Zeigel RF, Twiehaus MJ (1966) Biological studies with RE virus (strain T) that induces reticuloendotheliosis in turkeys, chickens, and Japanese quails. J Natl Cancer Inst 37:731-743

Wilhelmsen KC, Eggleton K, Temin HM (1984) Nucleic acid sequences of the oncogene v-*rel* in reticuloendotheliosis virus strain T and its cellular homolog, the proto-oncogene c-*rel*. J Virol 52:172-182

Zabel U, Henkel T, dos Santos Silva M, Baeuerle PA (1993) Nuclear uptake control of NF-κB by MAD-3, an IκB protein present in the nucleus. EMBO J 12:201-211

Subject Index

Printing: Saladruck, Berlin
Binding: Buchbinderei Lüderitz & Bauer, Berlin